Innovations in Sustainable Technologies and Computing

The book series aims to publish research on the analysis and development of technological innovations that consider natural resources and nurture economic and social development. Innovative sustainable technologies are expected to develop sustainable production planning and tools, which reduces environmental and ecological risks drastically.

The series covers research related to innovative solutions in the field of sustainable technology, computing, and communication. Computational methodologies in the field of computer science and engineering, cybersecurity, data science, information systems and software engineering, algorithms for communication, smart transport system, smart city planning, e-waste management system, and other such sustainable technological solutions within the scope of this series.

The series will publish monographs, edited volumes, textbooks and proceedings of important conferences, symposia and meetings in the field of sustainable technology and computing.

Makarand R. Velankar · Parikshit N. Mahalle ·
Gitanjali R. Shinde

Cognitive Computing for Machine Thinking

Springer

Makarand R. Velankar
Department of Information Technology
MKSSS's Cummins College
of Engineering for Women
Pune, Maharashtra, India

Gitanjali R. Shinde
Department of Computer Science
and Engineering (AI and ML)
Vishwakarma Institute of Information
Technology
Pune, Maharashtra, India

Parikshit N. Mahalle
Department of Artificial Intelligence
and Data Science
Vishwakarma Institute of Information
Technology
Pune, Maharashtra, India

ISSN 2731-880X ISSN 2731-8818 (electronic)
Innovations in Sustainable Technologies and Computing
ISBN 978-981-97-0454-5 ISBN 978-981-97-0452-1 (eBook)
https://doi.org/10.1007/978-981-97-0452-1

This Springer imprint is published by the registered company Springer Nature Singapore Pte Ltd.
The registered company address is: 152 Beach Road, #21-01/04 Gateway East, Singapore 189721, Singapore

Paper in this product is recyclable.

Preface

The mind is everything. What you think you become.
—Bhagvad Gita

Cognitive computing has grabbed the attention of researchers recently. It deals with the human mind and thinking to make appropriate decisions considering human psychology. The current AI models are mainly based on data-driven models and require huge data to train the models. The issues that need to be tackled by AI models include fairness, deviation from intended objectives, ethical principles, sensitive information and security, performance disparity, privacy, and copyrights. Challenge is to align the AI models to perform in the right manner especially to the critical decision-making systems such as legal or medical applications. An augmented intelligent model with embedded cognitive computing and machine thinking paradigm can be a possible approach to tackle the current limitations of AI and advance the research in computing domain. Machine thinking is the new paradigm shift proposed in this book. The challenges in machine thinking are understanding and modeling the cognitive thinking process of humans that are discussed. The proposed paradigm will be useful in improving AI application performance which involves human cognition.

This book fundamentally aims at presenting cognitive modeling along with the new paradigm machine thinking to enhance existing AI power and address its current limitations. This book provides an overview of natural and artificial intelligence along with the computing models used currently. The need to advance the current models is presented with suitable examples. The business case studies presented in different domains provide possible use of augmented intelligence with the proposed machine thinking paradigm.

This book is targeted at academicians, researchers, students, and professionals who belong to disciplines which involve intelligent computing and modeling human thinking. It provides possible multidisciplinary research directions including social psychology, artificial intelligence, human–computer interaction, and cognition for applications in various domains.

There is a growing interest in the field of artificial intelligence (AI) and machine learning, and with that interest comes a need for a better understanding of how machines think. A book on machine thinking could be an important resource for individuals who want to learn more about this field, including researchers, engineers, and anyone who is interested in the potential applications of AI and machine learning. This book helps individuals understand the capabilities and limitations of machines, including how they process information and make decisions. This knowledge could be crucial for individuals who are working on developing new AI systems or who are incorporating AI into their businesses or organizations. This book also explores the future of this rapidly evolving field. As new technologies emerge and new applications for AI are developed, it is important for individuals to stay up to date on the latest developments and to be prepared for the changes that lie ahead. A book on machine thinking could be an important resource for individuals who want to stay informed and engaged with this dynamic field.

The main characteristics of this book are:

- Provides a strong foundation for the readers regarding the AI models, their advantages, and limitations.
- Provides illustrative examples for better understanding and applicability.
- Elaborates on the need of augmented intelligence with human cognition.
- Equips the readers with the ability to identify issues at hand (specifically regarding artificial intelligence) thereby making the users understand the need of new paradigm as machine thinking.
- Provides in-depth understanding on the various machine learning models used for end-to-end applications.
- Provides numerous case studies for providing a clear understanding of concepts in real time and serves as a guide to aspiring students/professionals by presenting various research openings in the field.
- Simple and easy language so that it can be useful to a wide range of stakeholders like a layman to educate users, villages to metros, and national to global levels.

This book is useful to all undergraduate students of computer science and AI courses for better understanding of creative AI, project development, and product design in machine thinking and artificial intelligence. This book is also useful to a wider range of researchers and design engineers who are concerned with exploring machine thinking for engineering use cases. Essentially, this book is most useful to all entrepreneurs who are interested in starting their start-ups in the field of application of machine learning and machine thinking to civil, mechanical, chemical engineering, and healthcare domain as well as related product development. The book is useful for undergraduates, postgraduates, industry, researchers, and research scholars in ICT, and we are sure that this book will be well-received by all stakeholders.

Pune, India

Dr. Makarand R. Velankar
Dr. Parikshit N. Mahalle
Dr. Gitanjali R. Shinde

Contents

About the Authors

Dr. Makarand R. Velankar received Ph.D. degree in Computer Engineering from Savitribai Phule Pune University. He has 11 years of industry experience and is associated with MKSSS'S Cummins College of Engineering for women, Pune, Maharashtra, India since 2001. His research interests include machine learning, AI, information retrieval, algorithms, cognitive computing and computational musicology. He is member of various professional bodies such as Computer Society of India, ACM, Indian Institute of Engineers. He acted as reviewer for many journals and conferences such as IEEE transactions, Springer, Eurasip, Sadhna, UAAI, IGI Global and ISMIR. He is associated with International Society of Music Information Retrieval to promote research in computational music. He is also an entrepreneurship educator, trainer and approved mentor for start-ups. He received Gold Global Jury award for evaluating and guiding startups at global level. He has delivered talks and conducted sessions/webinars on different topics in startups, entrepreneurship and computing.

Dr. Parikshit N. Mahalle is a senior member IEEE and is Professor, Dean Research and Development and Head—Department of Artificial Intelligence and Data Science at Vishwakarma Institute of Information Technology, Pune, India. He completed his Ph.D. from Aalborg University, Denmark and continued as Post Doc Researcher at CMI, Copenhagen, Denmark. He has 23+ years of teaching and research experience. He is an ex-member of the Board of Studies in Computer Engineering, Ex-Chairman Information Technology, Savitribai Phule Pune University and various Universities and autonomous colleges across India. He has 15 patents, 200+ research publications (Google Scholar citations-2900 plus, H index-25 and Scopus Citations are 1400 plus with H index-18, Web of Science citations are 438 with H index-10) and authored/edited 54 books with Springer, CRC Press, Cambridge University Press, etc. He is editor in chief for *IGI Global—International Journal of Rough Sets and Data Analysis, Inter-science International Journal of Grid and Utility Computing*, member-Editorial Review Board for *IGI Global—International Journal of Ambient Computing and Intelligence* and reviewer for various journals and conferences of the repute. His research interests are Machine Learning, Data Science, Algorithms, Internet of Things, Identity Management and Security. He is guiding 8 Ph.D. students in the area of IoT and machine learning and SIX students have successfully defended their Ph.D. under his supervision from SPPU. He is also the recipient of "Best Faculty Award" by Sinhgad Institutes and Cognizant Technologies Solutions. He has delivered 200 plus lectures at national and international level.

Dr. Gitanjali R. Shinde has overall 15 years of experience, presently working as Head and Associate Professor in Department of Computer Science and Engineering (AI and ML), Vishwakarma Institute of Information Technology, Pune, India. She has done Ph.D. in Wireless Communication from CMI, Aalborg University, Copenhagen, Denmark on Research Problem Statement "Cluster Framework for Internet of People, Things and Services"—Ph.D. awarded on 8th May 2018. She obtained M.E. (Computer Engineering) degree from the University of Pune, Pune in 2012 and B.E. (Computer Engineering) degree from the University of Pune, Pune in 2006. She has received research funding for the project "Lightweight group authentication for IoT" by SPPU, Pune. She has presented a research article in the World Wireless Research Forum (WWRF) meeting, Beijing China. She has published 50+ papers in National, International conferences and journals. She is author of 10+ books with publishers Springer and CRC Taylor & Francis Group and she is also editor of books. Her book *Data Analytics for Pandemics: A COVID–19 Case Study* is awarded outstanding Book of year 2020.

Chapter 1
Natural and Artificial Intelligence: Overview

1.1 Natural Intelligence

The need for intelligence is becoming an integral part of every use case surrounding us. Due to increasing demand for intelligent automation and digitization, intelligence has become the mandatory component of a plethora of applications [1]. The term "intelligence" has various meanings based on the context it is being referred to. However, the dictionary meaning of word intelligence refers to the ability to learn, understand, reason, and make judgments or have opinions that are based on reason. It is the capacity for abstraction, logic, self-awareness, emotional knowledge, learning, and reasoning. Intelligence can be defined in various ways, and it includes problem-solving and decision-making skills that mimic the capabilities of the human mind. It is the ability to think, reason, and understand instead of doing things automatically or by instinct. Practically, there are three types of intelligence, viz. real intelligence, natural intelligence, and the most popular buzzword, i.e., artificial intelligence (AI). Before we get into the details of AI and its applications, it is very important to understand these two types of intelligence in the context of similarities and differences.

Real intelligence is the first type of intelligence in hierarchy which deals with reality in the context of problem-solving. Few of the examples include world models, data, information, and knowledge representation for the key purposes like cognition, learning, and reasoning. The main outcomes of real intelligence are decision-making, understanding, and problem-solving while interacting with the environment. Natural intelligence [2, 3] refers to the intelligence that is created by nature, natural evolutionary mechanisms, as biological intelligence embodied as the brain, animal, etc. It is an innate cognitive ability that humans and other animals are born with, which allows them to learn from their environment, experience, and interactions. It is the ability to perceive, understand, learn, and solve problem using cognitive and reasoning abilities. Natural intelligence is a product of evolution and is embodied in the biological structures of the brain. It encompasses a wide range of mental processes, including

M. R. Velankar et al., *Cognitive Computing for Machine Thinking*, Innovations in Sustainable Technologies and Computing, https://doi.org/10.1007/978-981-97-0452-1_1

perception, memory, attention, language, and creativity. Natural intelligence allows organisms to interact with their environment, adapt to changes, and navigate complex situations.

Natural intelligence involves various processes and mechanisms that allow organisms to perceive, understand, learn, and solve problem in their environment. While the intricacies of natural intelligence are still being explored, here are some key aspects of how it works are discussed below:

1. **Neural Networks**: Natural intelligence relies on complex networks of neurons in the brain. These neurons communicate with each other through electrical and chemical signals, forming neural networks that process and transmit information.
2. **Perception**: Natural intelligence involves the ability to perceive and interpret sensory information from the environment. This includes the senses of sight, hearing, touch, taste, and smell. Perception allows organisms to gather information about the world and make sense of it.
3. **Learning and Memory**: Natural intelligence encompasses the capacity to acquire knowledge and skills through learning. Learning involves the process of acquiring and modifying information, behaviors, and concepts based on experience. Memory allows organisms to store and retrieve information, facilitating learning and decision-making.
4. **Reasoning and Problem-Solving**: Natural intelligence enables organisms to reason, think logically, and solve problems. It involves processes such as deductive reasoning, inductive reasoning, critical thinking, and problem-solving strategies. These cognitive abilities help organisms make decisions and adapt to new situations.
5. **Language and Communication**: Natural intelligence includes the ability to use language and communicate with others. Language allows for the expression of thoughts, ideas, and emotions, facilitating social interaction and the exchange of information.
6. **Attention and Focus**: Natural intelligence involves the ability to selectively attend to relevant information while filtering out distractions. Attention allows organisms to focus on specific stimuli or tasks, enhancing perception, learning, and problem-solving.
7. **Emotion and Motivation**: Natural intelligence is intertwined with emotions and motivation. Emotions play a role in decision-making, memory formation, and social interactions. Motivation drives behavior and goal-directed actions, influencing learning and problem-solving.
8. **Adaptability and Flexibility**: Natural intelligence enables organisms to adapt to changing environments and circumstances. It involves the ability to learn from mistakes, adjust behaviors, and explore new strategies when faced with challenges.

However, it is distinct from artificial intelligence (AI), which is the intelligence demonstrated by machines or computer programs. Natural intelligence has been a subject of study in various fields, including cognitive science [3], neuroscience, and psychology. There are also companies that have adopted the name and have focused on intent marketing or developing AI models inspired by natural intelligence. AI [4] is the simulation of human intelligence in machines that are programmed to think and learn like humans. It involves the development of computer algorithms and systems that can perform tasks that require human-like intelligence, such as understanding natural language, recognizing images and patterns, making decisions, and solving problems. AI encompasses a wide range of subfields, such as machine learning, natural language processing, computer vision, robotics, and expert systems, among others. AI has numerous practical applications across various industries, including healthcare, finance, transportation, and manufacturing, among others [5].

AI has become increasingly important because it has the potential to revolutionize the way we live and work. Some of the key reasons why we need AI include [6]:

- **Enhancing efficiency and productivity**: AI can automate routine and repetitive tasks, freeing up human workers to focus on more strategic and high-level tasks. This can help companies increase efficiency and productivity and reduce costs.
- **Improving decision-making**: AI can analyze large amounts of data and provide insights that can help humans make better decisions. This can be particularly useful in areas such as healthcare, finance, and marketing.
- **Enabling new products and services**: AI can enable the development of new products and services that were previously impossible or difficult to create. For example, AI-powered virtual assistants and chatbots can provide personalized customer service, while self-driving cars can improve transportation and reduce traffic accidents.
- **Addressing societal challenges**: AI can also be used to address some of the world's most pressing societal challenges, such as climate change, poverty, and healthcare. For example, AI can help optimize energy use, improve access to healthcare in remote areas, and support disaster relief efforts.

AI has the potential to transform many aspects of our lives, from the way we work to the way we live and interact with each other. As such, it is important to invest in AI research and development and to ensure that AI is used in a responsible and ethical manner. There are several types of artificial intelligence (AI), which can be classified in various ways depending on the criteria used. Some common types of AI include [7]:

1. **Reactive Machines**: These are AI systems that can react to and interact with their environment but do not have the ability to learn or store past experiences. Examples include Deep Blue, the chess-playing computer, and Siri, the virtual assistant.
2. **Limited Memory**: These are AI systems that can learn and store past experiences, but only for a limited amount of time. Examples include self-driving cars, which can learn from past driving experiences to make better decisions in the future.
3. **Theory of Mind**: These are AI systems that can understand and interpret the emotions, beliefs, and intentions of other entities. This is an important area of research for developing more human-like AI systems.

4. **Self-aware**: These are AI systems that have a sense of self and consciousness and can understand their own existence and limitations. This type of AI is currently the subject of much debate and speculation and is largely the realm of science fiction.
5. **Narrow AI**: This is a type of AI that is designed to perform specific tasks, such as playing chess or recognizing speech. Most of the current AI systems fall under this category.
6. **General AI**: This is a type of AI that can perform any intellectual task that a human can do, such as learning, reasoning, and problem-solving. This type of AI is still largely theoretical, and much research is being done to develop more advanced systems.
7. **Super AI**: This is a hypothetical type of AI that is vastly more intelligent than any human and is capable of solving problems and making decisions beyond the scope of human understanding. This type of AI is also largely the realm of science fiction and is the subject of much speculation and debate.

AI is progressing rapidly. AI progression refers to the advances, milestones, and breakthroughs that have been achieved in the field. Over the years, there has been significant progress in AI research and development, resulting in increasingly sophisticated AI systems that can perform a wide range of tasks. Some of the key areas of progress in AI include machine learning, natural language processing, computer vision, robotics, and expert systems, among others. The development of more powerful computing and communication capabilities, better data infrastructure, and the availability of large datasets have all contributed to the progress in AI. Some of the potential benefits that AI progression could bring include increased efficiency and productivity, improved decision-making, and the development of new products and services. At the same time, there are concerns about the potential risks and challenges associated with AI progression, such as job displacement, ethical considerations, and the potential for AI systems to be misused or abused.

Human reasoning [8] refers to the cognitive processes and capabilities that allow humans to think, make decisions, and solve problems. It encompasses a broad range of mental processes, including perception, judgment, inference, logic, and creativity. Human reasoning is connected to many aspects of cognition [9, 10], including memory, attention, and language, and is shaped by individual experiences and the cultural and social context in which a person lives. Research in psychology and cognitive science has shed light on many aspects of human reasoning, including the cognitive biases and heuristics that can influence decision-making and problem-solving. Advances in artificial intelligence are also contributing to our understanding of human reasoning and the development of machine-based reasoning systems that can simulate some aspects of human thought and decision-making.

1.2 Human-like Reasoning and Cognition

Human-like reasoning [11] is an area of research in artificial intelligence (AI) that aims to develop AI systems that can reason and make decisions in a way that is similar to human beings. This involves incorporating aspects of human cognition, such as intuition, common sense, and creativity, into AI systems. Some of the current approaches to achieving human-like reasoning in AI include using machine learning techniques such as deep learning and reinforcement learning, as well as incorporating knowledge representation, natural language processing, computer vision, and other areas of AI research. The potential benefits of developing AI systems with human-like reasoning capabilities include improved decision-making, enhanced accuracy, and the ability to handle complex and uncertain situations. However, there are also concerns about the potential risks and challenges associated with developing such advanced AI systems, such as the potential for bias, ethical considerations, and the loss of human jobs to AI automation.

Decision-making refers to the process of identifying, gathering, and assessing information and selecting a course of action or belief among several alternatives. It is a cognitive process that involves the use of reasoning, judgment, intuition, and other mental processes. Effective decision-making requires the ability to weigh evidence, consider the context, and anticipate the consequences of different choices. Decision-making can be a complex and challenging process, particularly when faced with uncertainty, conflicting goals, or limited resources. There are various theories and models of decision-making, and the process can vary depending on the specific situation and the individual or group involved. Examples of decision-making include choosing a career path, deciding on a course of treatment for a medical condition, or selecting a product or service to purchase. Decision-making and AI are closely related, as AI has the potential to significantly improve decision-making in a wide range of contexts. AI technologies, particularly machine learning, can help identify patterns and insights in large amounts of data, allowing individuals and organizations to make more informed decisions. AI-powered decision-making tools can help with tasks such as risk assessment, forecasting, and resource allocation and can provide real-time recommendations and insights. The use of AI in decision-making can also help reduce bias and improve accuracy, particularly when dealing with complex datasets. However, there are also concerns about the potential risks and challenges associated with relying too heavily on AI in decision-making processes, such as the potential for errors or malfunctions, the need for ethical considerations, and the potential impact on jobs and human decision-making.

Cognition [12] refers to the mental processes involved in knowing, including perception, thought, judgment, reasoning, learning, and problem-solving. It encompasses both conscious and unconscious processes and involves the acquisition and use of knowledge and understanding through sensory experience, memory, perception, and other cognitive functions. Cognition is a complex and multidisciplinary area of study that draws on aspects of psychology, neuroscience, philosophy, and computer science, among other fields. Similarly, cognition and AI are both related to the study of how the mind works. Cognition refers specifically to the mental processes involved in knowing, including perception, thought, judgment, reasoning, learning, and problem-solving. AI, on the other hand, refers to the development of machines and systems that can perform tasks that typically require human intelligence, such as perception, decision-making, and language processing. There is significant overlap between cognition and AI, as researchers in both fields are interested in understanding how the mind works and how it can be replicated or emulated by machines. Research in AI often draws on insights from cognitive science and neuroscience, while advances in AI technology can inform our understanding of cognition and its limitations.

1.3 Artificial Intelligence

Machine learning [13] has been a prominent technique over the past two decades in the area where there is requirement of decision, learning, prediction, remember, analyze, and recognize. There are a plethora of applications and use cases using machine learning for the requirement mentioned above. However, there is a tradeoff between the performance of machine learning algorithms and the data size. It should be noted that with the increasing size of dataset, machine learning performance also increases. However, after a certain size of the dataset, machine learning algorithms start performing constant, or in other words, these algorithms start underperforming. This tradeoff is depicted in Fig. 1.1.

This tradeoff [14] indicates that depending on the underlined application, after a certain size of the dataset deep learning becomes a better choice for processing. This choice of machine learning or deep learning algorithms always varies based on the application and set of questions to be posted on a dataset for the purpose of business

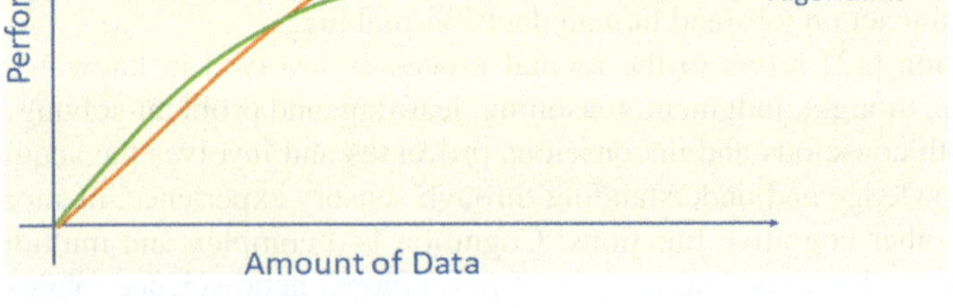

Fig. 1.1 Scaling with amount of data [14]

intelligence. Each technique consists of a rich set of algorithms which includes classification and clustering in machine learning and identification and detection in deep learning. Literature shows that these algorithms in each category have application specific limitations.

Some of the limitations of machine learning algorithms include overfitting and under-fitting, lack of data and interpretability, ethical concerns, deterministic problems, inaccuracy, high error chances, and deep learning algorithms requiring a large amount of data. However, it is worth noting that machine learning is a powerful form of artificial intelligence that has the potential to observe trends, patterns, and behaviors. Additionally, machine learning algorithms have been shown to have high diagnostic accuracy in detecting diseases such as diabetic retinopathy. Understanding the limitations of machine learning is important in order to use these algorithms effectively and responsibly.

In addition to this, the limitations of deep learning algorithms include high computational cost, overfitting, and lack of interpretability, dependence on data quality, data privacy and security concerns, lack of domain expertise, unforeseen consequences, and limited applicability to certain types of problems. Additionally, while deep learning has the ability to handle large and complex data and achieve state-of-the-art performance, it also requires a large amount of data and may suffer from lack of generalization if the data is not representative of the problem as well as limitation of specific activation function used. Understanding the limitations of deep learning is important when applying it to a specific problem and developing effective solutions.

It is very important to note that the machine learning and deep learning algorithms cannot think in the same way that humans do. They are programmed to learn and make decisions based on patterns and statistical analysis of data, but they are not capable of true cognitive thought or consciousness. Instead, they rely on the data and algorithms provided to them to make decisions and provide outputs. Deep learning may not perform well in cases where there is insufficient labeled data or where the data is heavily imbalanced. Additionally, machine learning and deep learning algorithms may not be appropriate for tasks that involve complex decision-making processes or reasoning tasks that require human-level cognitive abilities. There may be certain tasks or domains where machine learning and deep learning cannot perform effectively [15]. This includes:

- Tasks that involve complex decision-making processes or reasoning tasks that require human-level cognitive abilities.
- Soft skills such as emotional intelligence and reasoning abilities that are critical for workplace development and growth.
- Tasks that require human labeling and categorization of data for training the AI models.
- Situations where these techniques are unable to function optimally without relying on innately human factors such as intuition, creativity, and judgment.
- Situations where the data is insufficient or heavily imbalanced, leading to poor performance of machine learning or deep learning models.
- Tasks that operate beyond the defined functions of narrow AI models.

It is important to note that AI is designed to complement human intelligence and ability, rather than compete with it.

1.4 Model-Centric and Data-Centric AI

Model building was considered as one of the important steps in end-to-end AI application development. Model-centric AI [16] is a type of approach in which the primary goal is to produce the best machine learning model for a given dataset. In this approach, the focus is on selecting the best machine learning algorithms, programming languages, and AI platforms to create the best model possible. Optimizing and improving the performance of the model is the main objective. This approach treats data as a fixed artifact and focuses more on improving the algorithms. Several organizations have traditionally used this approach in developing AI systems. The key steps to be followed in model-centric AI are listed below:

1. **Data cleaning and preparation**: In this step, the dataset is prepared by removing any irrelevant or missing data and ensuring that the data is in the correct format for machine learning model training.
2. **Model selection**: The suitable model architecture and algorithm are selected for the given task.
3. **Model training and optimization**: The selected model is trained using the training data and optimized using various techniques such as hyperparameter tuning.
4. **Model testing and validation**: The performance of the model is evaluated using a separate test dataset and validated with various statistical metrics such as accuracy, precision, and recall.
5. **Deployment**: Finally, the model is deployed in a production environment where it can be used to make predictions on new data.

However, these steps may vary depending on the specific dataset and task and can be iterative in nature, with multiple rounds of training and testing required to achieve the desired results. Over the period of time, the amount of data is growing exponentially and model building is getting more mature. In the present scenario, no more efforts are required in building the machine learning or deep learning model [16] as the ready-made models are available on many online sources. Few limitations of model-centric AI are listed below:

- **Limitations to progress**: Model-centric AI requires that the dataset fits the model, meaning that the learning process is limited to the availability and quality of the training data.
- **Less reliable outcomes**: Because model-centric AI focuses more on improving the performance of the algorithms than on the quality of the underlying data, it may result in models that are brittle and struggle to generalize well to new data.

- **High costs**: Building and maintaining machine learning models can be costly for an organization as it requires specialist data science and computational resources.
- **Lack of training data**: High-quality labeled data is necessary to train machine learning models. However, obtaining training data can be challenging in certain situations, giving rise to deficiencies in the models produced.

To address these limitations, alternative approach such as data-centric AI is being explored and discussed further.

Data-centric AI [17, 18] is an emerging approach in AI development that prioritizes improving the quality and consistency of data used to train machine learning models, compared to the traditional model-centric approach which focuses primarily on model performance. In this approach, the quality of the training data is systematically enhanced to improve the accuracy and generalizability of the model. Data-centric AI emphasizes the importance of high-quality, labeled training data, and the quality of the data determines the performance of the AI system. Data-centric AI aims to create more reliable, scalable, and adaptable AI systems that can handle a wide range of real-world scenarios. The key steps of data-centric AI are listed below [18]:

- **Data collection and curation**: Gathering and selecting high-quality, relevant data is essential for building accurate models.
- **Data preparation and cleaning**: Data preparation involves manipulating and preparing the data, including cleaning and preparing it for analysis.
- **Data labeling and annotation**: High-quality labeled data is critical to building accurate models, and labeling it using appropriate annotations in line with the problem at hand is essential.
- **Data analysis and feature engineering**: Preparing the data so that it is ready for ML modeling by performing exploratory data analysis on the data and feature engineering to generate better predictive features.
- **Model selection and design**: Selecting the appropriate modeling architecture to suit the problem and creating a model that can generate accurate results.
- **Model training and optimization**: Training the model on the selected data and optimizing its performance using algorithmic tuning, hyperparameter optimization, and other methods.
- **Model testing and deployment**: Finally, testing the model on new data to ensure it can generalize well beyond the initial training dataset and deploying the model in a production environment.

These stages are iterative in nature, and the data-centric AI approach requires a flexible and adaptive methodology to improve and refine the model as new data becomes available. However, it is still a matter of exploration and debate whether the model-centric or data-centric AI approach is more suited for human-like reasoning and machine thinking.

Data-centric AI [19] is particularly relevant in fields where data is heterogenous or when there is a need to address biases, improve interpretability, or deal with limited data. It emphasizes the importance of understanding the data, its limitations, and potential biases and working toward creating reliable and representative datasets.

1.5 Summary

AI has been a phenomenal revolution over the last two decades, and the notion of AI has completely changed to data mining. The main fuel for each AI application is data and how quality data we provide decides the success of any AI model. Emergence of real intelligence to natural intelligence and in turn to artificial intelligence clearly indicated that there is a need to analyze different techniques of AI. This chapter mainly focuses on the tradeoff between machine learning and deep learning with respect to the performance and put forth key limitations of these popular techniques. This chapter also focuses on the need of machine thinking in the coming years in line with cognition and human-like reasoning. Finally this chapter concludes with the advantages and limitations of model-centric AI and data-centric AI. In fact, this chapter also presents the debatable point of which AI model out of this would be a better match for machine thinking.

References

1. Dey N, Wagh S, Mahalle PN, Pathan MS (2019) Applied machine learning for smart data analysis (1st ed). CRC Press/Taylor and Francis Group, New York, NY
2. Weng J (2013) Natural and artificial intelligence. BMI Press
3. Weng J (2012) Natural and artificial intelligence: introduction to computational BrainMind. BMI Press, Okemos, Michigan
4. Vernon D (2014) Artificial cognitive systems: a primer. The MIT Press
5. Borana J, Jodhpur NU (2016) Applications of artificial intelligence and associated technologies. In: Proceeding of international conference on emerging technologies in engineering, biomedical, management and science [ETEBMS-2016]
6. Mahalle PN, Gitanjali RS, Shinde GR, Pise PD, Deshmukh JY, Jyoti YD (2022) Data collection and preparation. In: Foundations of data science for engineering problem solving. Springer, Singapore, pp 15–31
7. Srinivasan R, Chander A (2021) Biases in AI systems Commun. ACM 64(8), 44–49
8. Johnson-Laird PN, Khemlani SS, Goodwin GP (2015) Logic, probability, and human reasoning. Trends Cogn Sci 19(4):201–214
9. Cassenti DN, Veksler VD, Ritter FE (2022) Editor's review and introduction: cognition-inspired artificial intelligence. Top Cogn Sci 14(4):652–664
10. Schmid U (2008) Cognition and AI. KI 08/1, Themenheft "Kognition", pp 5–7
11. Megill J (2014) Emotion, cognition and artificial intelligence. Minds Mach 24(2):189–199, Springer
12. Pollack M (2005) Intelligent technology for an aging population: the use of ai to assist elders with cognitive impairment. AI Mag 26(2):9–24
13. Khadse VM, Mahalle PN, Shinde GR (2020) Statistical study of machine learning algorithms using parametric and non-parametric tests: a comparative analysis and recommendations. Int J Ambient Comput Intell 11:80–105

14. Janiesch C, Zschech P, Heinrich K (2021) Machine learning and deep learning. Electron Markets 31:685–695, Springer
15. Sharifani K, Amini M, Machine learning and deep learning: a review of methods and applications. World Inf Technol Eng J 10:3897–3904
16. Hamid OH (2023) Data-centric and model-centric ai: twin drivers of compact and robust industry 4.0 solutions. Appl Sci 13:2753
17. Zha D, Bhat ZP, Lai KH, Yang F, Hu X (2023) Data-centric ai: perspectives and challenges. arXiv 2023, arXiv:2301.04819
18. Zha D, Bhat ZP, Lai KH, Yang F, Jiang Z, Zhong S, Hu X (2023) Data-centric artificial intelligence: a survey. arXiv 2023, arXiv:2303.10158
19. Zigon J (2019) Can machines be ethical? on the necessity of relational ethics and empathic attunement for data-centric technologies. Soc Res Int Q 86:1001–1022

Chapter 2
Cognitive Computing and Augmented Intelligence

2.1 Cognitive Computing

Human cognition is the study of the mental processes underlying our capacity to observe the surrounding, recall, represent, and learn from our experiences, and adjust our behavior accordingly. It comprises topics such as perception, attention, memory, knowledge, language, and the thought process about problem-solving, reasoning, and decision-making. The aspects of intelligence, emotions, and consciousness are integral parts of thinking. According to Ulric Neisser [1] whose work helped define cognitive psychology as a discipline, "cognition is all the processes by which the sensory input is transferred, reduced, elaborated, stored, recovered and used".

Cognitive psychology typically has some basic assumptions about the mind [2].

- The mind is considered as a data processing system which helps us to interact with the environment.
- The mind has limited resources, and the behavior can be generalized by studying behavioral aspects of humans.
- Human behavior is the result of external perception and existing knowledge.

Many theories about the cognition processes are developed by psychologists. One theory proposes that the cognitive processes are clusters of functions which have independent processing capabilities of specific info from sensory inputs. As an example, the visual aspects such as object recognition are processed independently, or speech recognition is an independent process based on the auditory sensory system. This has led to development of computing models based on neural networks such as convolutional neural networks for image processing and recurrent neural networks for time series data such as speech.

Human cognition can be represented as the function of input and output as shown in Fig. 2.1, where input is the sensory input and output is the behavior. Cognitive computing attempts to model this process using the existing knowledge about the brain and mind. The view with brain and mind is identical and led to the study

M. R. Velankar et al., *Cognitive Computing for Machine Thinking*, Innovations in Sustainable Technologies and Computing, https://doi.org/10.1007/978-981-97-0452-1_2

of neural activity in the brain. The view with brain and mind is different and are interrelated, attempting to find answers with explanations of behavior in terms of information processing and mental functions.

Marr [3] proposed a 3-level theory of analysis of cognitive systems: computational theory, specific algorithms, and physical implementation. Computational level refers to what to be computed along with why to be computed, i.e., problem description with the contextual info. Algorithm level refers to input representation, logic building to transform input into output to solve a particular problem. Implementation level provides details of algorithm execution plan to solve a particular problem. These levels were also referred at abstract levels as semantic, syntactic, physical [4] and content, form, medium [5], respectively.

Researchers have developed various theories for the computational theory of mind. The neural activity can be measured and used to explain cognitive behavior. The theory was further extended by Hilary Putnam (1967) and other cognitive scientists [6, 7] in recent years. Horst Steven [7] proposed "computational theory of mind" and posed questions like can a machine think? Or is the mind a thinking machine? Many cognitive scientists have proposed Turing-style models and computation for mental symbols. The representation theory of mind [8–10] presented by researchers discusses different viewpoints and theories about belief, symbol processing, language of thoughts, and concept formation. The theories proposed were challenged by other scientists and at times by the same scientists helped to build the foundation for cognitive computing.

In order to develop the cognitive computing model, one needs to understand the widely used terms in cognitive psychology. It includes systematic study of the human thinking process by developing the methods to validate the assumptions.

- **Phenomenon**: It is the behavioral aspect under consideration and study, i.e., the problem under consideration.
- **Theory**: It is the combination of assumptions and predictions which needs to be validated by experimental results.
- **Paradigm**: A set of theories and assumptions applicable to the phenomenon of interest.
- **Framework**: Guidelines or principles related to the phenomenon with boundaries.

Fig. 2.1 Human cognition representation

2.2 Phenomenon in Human Cognition

Human cognition involves various phenomena such as perception, attention, memory, mental representation, problem-solving, reasoning and decision-making, language, comprehension, intelligence, consciousness, and emotions. The introduction to these phenomena will provide a foundation for proposed machine thinking.

Perception and Attention

Sensory perception such as auditory and visual perception has achieved good results in terms of speech and object recognition with a deep learning approach. The internal representation of objects in the most meaningful manner for computing is a challenge as current systems are data driven. It does not include human angles such as selective attention and forgetting. Selective attention allows us to focus on specific aspects in the data to extract specific info while ignoring or less attention to other aspects. Our ability to hear our name despite the fact that we are not the part of the conservation and process different auditory input is an example of selective attention. Visual attention includes attention by particular objects or specific locations. Our ability to do multitasking with divided attention depends on tasks involved and practice. The regular tasks we perform such as driving on the same route require less attention once we are used to it.

Memory and Mental Representation

Human memory modeling is a serious challenge in cognitive computing. Humans typically have short-term memory or temporary memory and long-term memory or more permanent memory. It has various components and distinctions such as past or futuristic memory. Past memory involves semantic or knowledge and meanings and specific events with associations. Futuristic memory involves tasks to be done on a specific day or time. Another way to distinguish the memory is declarative and procedural. Declarative memory refers to remembering capitals of a country, whereas procedural memory refers to remembering the skills like playing any game. Recent memory distinctions are implicit and explicit. Implicit refers to our unconscious memories which we can recall due to specific events or reference, whereas explicit refers to conscious memories which we can report with more clarity. Memory is a crucial component as it influences our behavior significantly. Memory or mental representation is the result of the perception, attention, and interpretation, which is stored as info in our memory. It is attempted by storing features or abstract representations using machine learning for computing purposes. Various mental models are proposed by cognitive scientists to represent physical and abstract entities. Physical objects representation with structural features or spatial info or tokens/symbols and their interrelationships for the language are examples of mental representations. Concepts and their associations are represented as mind maps. Several scientists argued that different representations are interlinked and are used depending on the cognitive processes required to perform specific tasks or solve a particular problem.

Problem-Solving

State space model for problem-solving is widely used for problems in the domain such as gaming, robotics. Typically goals and states model is used to model a particular problem with start and end state. Expert systems developed using computer programming with domain knowledge and analogical reasoning solve specific problems efficiently. Problem-solving requires first defining and interpretation of the problem in hand. This involves framing and reframing the problem to arrive at the right representation form. The thinking process to arrive at right representation followed by selecting appropriate solutions based on knowledge, reasoning, and intelligent decision-making is quite challenging for computational modeling. At times our subconscious mind keeps thinking about the problem while we continue doing our regular work or while sleeping. The solution may be clicked with an "Aha" moment anytime, and it is also noticed by renowned scientists when they find an innovative solution.

Reasoning and Decision-Making

It was argued that mental models in abstract representation built based on premises result in the conclusions or decisions [11]. The errors may be possible due to limited working memory capacity, since all potential models must be compared for the conclusive decision. Another approach proposed was using mental logic having deductive and inductive reasoning with factual knowledge, and rational logic helps to arrive at a specific decision. Mental models or logic, both have challenges with limited working memory capacity to arrive at decision.

Language

Language in any form is the media to communicate the message or feelings. According to the linguistic theory proposed by Chomsky [12], the human brain has the ability to acquire language named as language acquisition device or LAD which is based on the hypothesis that all languages share a common universal grammar. It means the rules are inborn and further developed with the exposure to specific spoken or musical language. Second theory proposed is learning theory, which assumes that we learn from the exposure through reinforcement. Another theory as interactionist learning says social interactions and communication help to learn language [13]. Language and cognition have a strong interconnection with different viewpoints about the nature of connection. It is like the problem of thinking influences language or language influences thinking which can be considered as chicken or egg first problem. Thinking process is internal, whereas language is in the form of expression to share info.

Comprehension

Comprehension is the ability to decode the language in order to understand the meaning conveyed. Piaget [14] proposed a 4-stage theory of cognitive development. He argued that the concept should be first understood before acquiring the language to express it. The recent artificial intelligence application with a large language model named ChatGPT raises the question about how humans acquire the language.

ChatGPT or similar tools have shown excellent results; however in their current versions, they lack capabilities of humans such as common sense, multitasking, or creativity.

Intelligence
The theory of multiple intelligence [15] proposed different forms of intelligence. Various tests have been developed to measure different aspects of intelligence since the Binet test which has been used as a benchmark since its foundation of intelligence scale in 1909. Cattell proposed testing of cognitive abilities referred to as fluid intelligence and crystalized intelligence which were further modified by his student Horn with nine abilities. Carroll proposed a three-level model to measure abilities. Cattell–Horn–Carroll theory (CHC theory) which is a combined effort presented in 1999 influences the current tests significantly. CHC theory [16]: The relative importance of genetics and environment in intelligence building is a debatable issue.

Consciousness
Conscious term is used for the state of our mind such as conscious or fully awake and unconscious state. Sigmond Freud [17] the well-known psychologist proposed an additional state of mind between conscious and unconscious as a preconscious mind. Various theories and criticisms presented by researchers provided different viewpoints about states of mind and their interconnections. The cognitive processes in the conscious state may have some influence from the other states of mind according to researchers.

Emotions
Emotions are referred to as rasa by Bharat Muni in Vedas with Navarasa theory having nine states of mind. Similar emotion classification approaches with distinct emotions are proposed by researchers. Dimensional models such as the circumplex model by Russell [18] with valence and arousal axes classify the emotions in different manners. Emotions have various expressions and representations with multiple cultures that lead to different emotion taxonomies. Intensity of the emotions is another interesting aspect of measure. Emotions impact our thought process, behavior, and decisions. Emotions and cognition are strongly correlated with influence on each other.

These various phenomena of cognition make it a complex model for computing. Understanding and modeling these phenomena requires a multidisciplinary approach with contribution from various research streams.

2.3 Augmented Intelligence

Defining intelligence is itself a challenge, and various philosophers and psychologists have their own interpretations and definitions. Some definitions given by researchers are given to understand different viewpoints. According to Lloyd Humphreys, "It is the resultant of the process of acquiring, storing in memory, retrieving, combining, comparing, and using in new contexts information and conceptual skills". "The unique propensity of human beings to change or modify the structure of their cognitive functioning to adapt to the changing demands of a life situation" was proposed by Reuven Feuerstein.

According to artificial intelligence (AI) researchers, "Intelligence measures an agent's ability to achieve goals in a wide range of environments".

AI aims at making machines simulate and perform tasks like humans. In recent years, deep learning has shown promising results to perform certain tasks. The success of AI systems such as Deep Blue, an expert system to play chess, or AlphaGo for playing board games has shown a promising future. However, such systems are trained with the huge data and high computing resources. They lack their own thinking abilities like humans and are mainly data driven.

Considering the problems, uncertainty, and solutions provided by humans, it is argued that machines cannot totally replace humans. It is necessary to build a human cognitive-based AI system to improve performance of current AI systems, and it can help humans to understand some cognitive phenomena in a better manner. Human intelligence and AI have their own strengths and limitations. Humans perform better in some tasks such as learning under an uncertain or noisy environment with limited data or unquantifiable info and recognizing emotions compared with machines. Machines are better at tasks such as memorizing, computing, and working for long hours. Humans are better at doing tasks which are ill-defined, whereas machines are better at performing well-defined tasks. Combining both can help in overcoming the limitations of each other and improve strengths as a team.

Augmented intelligence is a human-centric model aimed at enhancing cognitive abilities with active alliance between machine and human. This can be achieved by embedding cognitive computing in machines in different ways such as human-influenced AI decisions, AI-influenced human decisions, and finally combined decision-making. A literature survey [19] about expert views on relations between human and machine intelligence and similar work provides directions for the development of augmented intelligence.

Different approaches for augmented intelligence are shown in Fig. 2.2. In the case of human-influenced AI, humans provide contextual info along with active feedback to AI for better decision-making. In the case of AI-influenced human systems, AI provides knowledge and inputs to humans as a supporting decision-making unit. In both these approaches, the results can be further fed into a system for improving decisions which are finally taken by AI or humans in respective models. In the case of a combined decision-making system, both AI and humans provide inputs and info to arrive at a common decision by evaluating different parameters as collaborative intelligence.

The augmented intelligence system needs to have certain characteristics such as rules and practical considerations for decision-making, efficient use of knowledge, and learning mechanism. Some important issues to be considered while building an intelligence are:

Fig. 2.2 Approaches for augmented intelligence

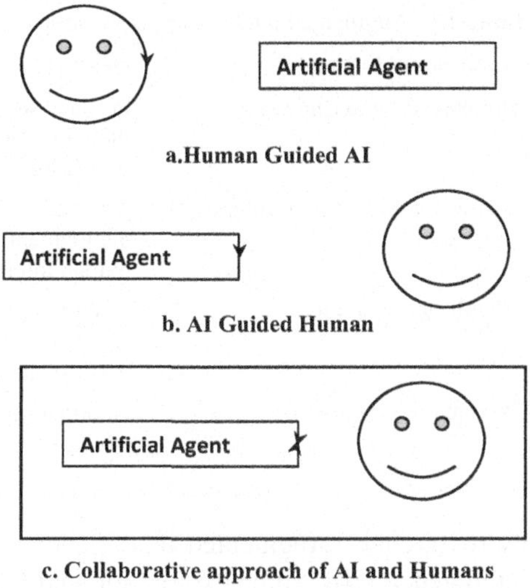

- Should the human brain structure be used as a template for design and implementation?
- Should human behavioral and cognitive capabilities be considered as a basis?
- How can thinking capabilities be embedded along with improved learning mechanisms?
- How to model uncertainty and cognitive capabilities?
- Can the system be generalized or have a special purpose for specific tasks?
- To use top down or bottom up or mixed approach for data processing, fusion, and analysis?
- How can the system learn and understand the context along with domain info?
- How to build mental models for decision-making?
- How to embed qualitative relative measures which are difficult to quantize?
- How to measure the intelligence of the proposed system?

From the representative questions mentioned, it clearly indicates that many issues remain open. Building a generalized augmented intelligence system needs to solve some basic questions about the expectations from it.

2.3.1 Applications of Augmented Intelligence

Various applications are possible in different domains with augmented intelligence, and some works in progress provide future possibilities with augmented intelligence.

Brain–computer interface with inputs to AI by humans based on sensory inputs and human decision-making process or inputs to humans by AI based on the data are shown as examples in sci-fi movies and proposed or partially experimented

Table 2.1 Augmented intelligence applications

Application/domain	Description	References
Human–computer interface	Produce suggestions based on human thought process considering the contextual info from environment	[21]
Games for cognitive development	A framework for rapid development and deployment for serious game building with human intelligence	[22]
Healthcare/hospitals	Collaboration in clinical environments and conversational agents to improve performance	[23]
Production systems	A human-guided exploration reinforcement strategy to maximize gain	[24]

by researchers. Augmented intelligence is also referred to as hybrid or collective intelligence [20]. Some exploration and research work are summarized as possible applications in Table 2.1.

The deployed examples such as Teachable Machines from Google with image recognition or Vencortex as a holistic system for business decisions show the possibilities and power of hybrid or augmented intelligence. There is a need for more research and work in the domain of augmented intelligence to make an impact in the race between AI and humans to develop a collaborative approach to solve the problems. This needs better understanding of behavior of humans and machines and possible exploration of them to work together as a team.

2.4 Machine and Human Behavior

2.4.1 Machine Behavior

Machine behavior or response is typically programmed or data driven with learning from the samples. The goals are set with specific expected performance. The study and understanding of machine behavior is important as it will help us to predict its impact on humans in different ways such as social, economic, cultural, and the overall growth along with its threats. AI-based systems and agents developed by scientists for doing specific tasks such as autonomous vehicles or drones or weapons, conversational systems such as Siri, intelligent chat programs, game playing programs, and automatic trading can have serious impact if they do not behave in the rational and expected manner.

Machine behavior is proposed as a scientific discipline to study considering its increasing importance and possible impact [25]. Various software systems are becoming part of our everyday life with increased use and reliance on them. Prediction of behavior and analysis of the complex algorithms working behind these systems is difficult or at times may not be possible considering the environment they operate in. The outcome of these systems may be harmful to the society, and this fear is a real threat. An algorithm-driven automatic justice system or the impact of manipulated social media on politics can have serious consequences. In case of an autonomous vehicle or weapon system, the wrong behavior may result in loss of human lives. In the financial sector, the algorithms developed should not lead to manipulation of markets such as crashing. The biased, manipulated, or malfunctioned AI can lead to disaster; thus it is necessary to analyze the machine behavior.

AI agents with automatic learning mechanisms learn and build algorithms based on the environment and exploration done. Relatively the codes of model description and training are simple compared to result analysis. With the given set of inputs and outputs, the system builds logic/process/algorithm which is hard to interpret. For the patented or proprietary systems, it is further difficult. A system is considered to be aligned if it surpasses the expected goals and is considered misaligned if it deviates from the desired intents. Once proven behavior is not sufficient, the further challenge is that the behavior should be aligned with the changing factors such as goals, environment, inclinations, and sensitive issues. Considering the severity of the issue, machine behavior is a critical concern and is likely to be a major challenge in the future. Machine behavior needs to be aligned with the intended objectives, preferences with ethical principles.

2.4.2 Human Behavior

It is being said that human behavior is unpredictable. It is true to some extent, but in the majority of cases the behavior is predictable. The behavior largely depends on the emotional state, education, brought up, psychology, experience, value systems, cultural aspects, age, exposure, wealth, maturity, contextual info, and the nature of the individual. In order to understand and predict human behavior, we need to understand social and psychological aspects of a group of individuals. Human behavioral data needs to be looked at from various angles to understand different perspectives and possible reasoning.

The behavioral aspects of our interest are mainly related to the tasks such as decision-making which involves intelligence, analytical thinking, and human cognition. Human behavior is used as training data for AI applications such as playing games or driving vehicles, but in such cases the machine is less likely to outperform human performance. In case a machine learns on its own with trial and error with huge instances, it may be able to beat humans such as games AlphaGo. The exploration by humans and capturing knowledge is far better than machines, and humans learn fast compared to machines in general. Humans can quickly learn to identify objects, whereas machines require a lot of samples or data to do the same task.

Human behavior can be tracked by various means and instances. Online behavior is tracked by many companies to provide recommendations. Human profile or persona can be created by the online interactions including use of search engines, social media, blogs, links, relationships, etc. Another approach is that based on specific context, we can analyze the behavior of individuals. The objective of understanding human behavior in a specific context may help in modeling the expected behavior of machines.

2.5 Summary

Cognition has roots in psychology and sociology and to model cognitive computing. It is essential to understand the basis of human cognition. The thought process of humans is influenced by many factors with prominence of sensory inputs one receives and the exposure with the environment. This chapter discusses the terminologies used in cognitive psychology and their relevance for building a computing model. Various phenomena of human cognition are discussed with the complexity levels they possess. Augmented intelligence is a combined effort of human intelligence and artificial intelligence. It is suitable to perform AI tasks involving cognitive abilities. Representative applications of augmented intelligence are presented and discussed. The need of understanding and predicting machine behavior is justified with possible serious consequences in critical applications. In order to model human cognitive capabilities and skills, it is necessary to understand and represent the knowledge acquired in the computable or quantitative form. This is one of the major challenges to develop augmented intelligent machines. Subsequent tasks include developing suitable models, algorithms, implementation, testing, and ethical issues involved. Prediction of machine and human behavior involves understanding the learning and knowledge acquired which can be explained to justify the decision-making process. Complementary skills of humans and machines together can be useful to build augmented intelligent machines in the future.

References

1. Neisser U (2014) Cognitive psychology: classic edition. Psychology Press
2. Andrade J, May J (2004) BIOS instant notes in cognitive psychology (1st ed). CRC Press
3. Marr D (1981) Vision: a computational investigation into the human representation and processing of visual information. MIT Press, Cambridge, Massachusetts
4. Pylyshyn ZW (1984) Computation and cognition
5. Glass AL, Holyoak KJ, Santa JL (1979) Cognition Addison-Wesley Publishing Company
6. Putnam H (1961) Brains and behavior
7. Horst S (2005) The computational theory of mind. In: The Stanford encyclopedia of philosophy
8. Fodor JA (1997) The representational theory of mind. Am Behav Sci 40(6):829–841
9. Sterelny K (1990) The representational theory of mind: an introduction. Basil Blackwell
10. Leslie AM (2000) How to acquire a 'representational theory of mind'. Metarepresentations: a multidisciplinary perspective, pp 197–223
11. Johnson-Laird PN (1983) Mental models: towards a cognitive science of language, inference, and consciousness (No. 6). Harvard University Press

12. Chomsky N (2006) Language and mind. Cambridge University Press
13. Vygotsky LS (2012) Thought and language. MIT press
14. Piaget J (2005) The psychology of intelligence. Routledge
15. Davis K, Christodoulou J, Seider S, Gardner HE (2011) The theory of multiple intelligences. The theory of multiple intelligences, in: Sternberg RJ, Kaufman SB (eds). Cambridge Handbook of Intelligence, pp 485–503
16. McGrew KS (2009) CHC theory and the human cognitive abilities project: standing on the shoulders of the giants of psychometric intelligence research. Intelligence 37(1):1–10
17. Freud S (1940) An outline of psycho-analysis. Int J Psychoanal 21:27–84
18. Russell J (1980) A circumplex model of affect. J Pers Soc Psychol 39(6):1161–1178
19. Yau KLA, Lee HJ, Chong YW, Ling MH, Syed AR, Wu C, Goh HG (2021) Augmented intelligence: surveys of literature and expert opinion to understand relations between human intelligence and artificial intelligence. IEEE Access 9:136744–136761
20. Dellermann D, Ebel P, Söllner M, Leimeister JM (2019) Hybrid intelligence. Bus Inf Syst Eng 61:637–643
21. Wang KJ, Zhang CY, Mao ZH (2019) Human-centered, ergonomic wearable device with computer vision augmented intelligence for VR multimodal human-smart home object interaction. In: Proceedings of 14th ACM/IEEE international conference human robot interactions. Daegu, South Korea, pp 767–768
22. Golestan S, Mahmoudi-Nejad A, Moradi H (2019) A framework for easier designs: augmented intelligence in serious games for cognitive development. IEEE Consum Electron Mag 8(1):19–24
23. Mirbabaie M, Stieglitz S, Frick NR (2021) Hybrid intelligence in hospitals: towards a research agenda for collaboration. Electron Mark 31:365–387
24. Doltsinis S, Ferreira P, Lohse N (2018) A symbiotic human machine learning approach for production ramp-up. IEEE Trans Human-Mach Syst 48(3):229240
25. Rahwan I, Cebrian M, Obradovich N, Bongard J, Bonnefon JF, Breazeal C, Wellman M et al (2019) Machine behaviour. Nature 568(7753):477–486

Chapter 3
Rethinking Machine Learning and Deep Learning

3.1 Mathematical Foundation for Machine Learning

Mathematics is the fundamental language of machine learning, understanding of mathematical concepts is needed for designing any machine learning algorithm. Algebra is the foundation of a lot of machine learning concepts and operations. Basically, it is all about using math symbols and equations. Linear algebra is important to understand how to represent and transform data. Vectors and matrices are used for coding features and relationships in data. Matrix operations like multiplication and inversion are also used in many machine learning algorithms like principal component analysis (PCA) and singular value decomposition (SVD). Linear algebra gives you insights into how much data is important and how much is not, which helps you reduce dimensionality and pick the right features. Algebraic concepts also help with optimization, where you can solve systems of equations to minimize loss functions. Having a good grounding in algebra is essential for machine learning, so you can represent and manipulate data well, create and implement algorithms, make sense of complex data, and optimize models for real-world problems.

Calculus is also one of the most fundamental branches of mathematics and plays an important role in machine learning. Calculus is at the core of machine learning because it provides us with the mathematical tools we need to understand how a function changes and how to optimize it. For example, in machine learning, we use derivatives or gradients to measure the change rate of loss functions. This allows us to fine-tune model parameters using optimization techniques such as gradient descent. We also use partial derivatives when dealing with a multivariate function, which is common in higher-dimensional data and in neural networks. Calculus also plays an important role when it comes to regularization techniques. For example, we can use L1 or L2 regularizations to prevent overfitting because we can add penalty terms to our loss functions. Having a good understanding of calculus is essential for machine learning because it helps us to train models efficiently, make better

M. R. Velankar et al., *Cognitive Computing for Machine Thinking*, Innovations in Sustainable Technologies and Computing, https://doi.org/10.1007/978-981-97-0452-1_3

model architectures decisions, and navigate complex landscapes of optimization and regularization.

The foundation of machine learning lies in probability and statistics. Probability theory provides the mathematical basis for understanding uncertainty, data-driven decision-making, and model performance. Machine learning practitioners use probability theory to model randomness in data, which is essential for tasks such as classification and prediction. Probability distributions, such as Gaussian and Bernoulli, as well as Poisson, help to describe and measure data patterns. Bayes' theorem serves as the foundation for Bayesian inference (BI), which is used in many machine learning algorithms to probabilistically model and make decisions. Statistics helps to extract meaningful insights from data, using concepts like expectation and variance to summarize data distributions. Maximum likelihood estimation and hypothesis testing help to estimate parameters and compare models, while cross-validation and other statistical techniques help to understand model predictions and assess model generalization.

3.2 Machine Learning Models

Machine learning is the process of teaching a computer to do something without being explicitly programmed. It is similar to teaching a child to do something by showing them examples and giving them feedback. Machine learning involves giving a computer large datasets and training it to recognize patterns and make predictions from that data. For instance, if you want a computer to recognize pictures of dog, you will show it many pictures of dogs and tell it to call them "dog". The computer will then learn common features in the images of dogs, like fur, ears, and eyes. Once it learns these features, it can identify new images that have never been seen before. Machine learning can be used in various domains, from predicting stock prices to diagnosing diseases. By providing a large amount of data to the computer, it can make very good predictions and in some cases outperform humans. The quality of the predictions depends on the data and the algorithm used to analyze it.

Machine learning is an artificial intelligence technique that does not require explicit programming to train a computer system to learn and improve from experience. In traditional computer programming, programmers write instructions for computers to follow, but with machine learning, computers analyze data and learn from it. This means that machine learning systems are able to make predictions based on patterns which are identified by learning data. Machine learning can be further classified as supervised, unsupervised, and reinforcement learning. Supervised learning involves training a machine learning model on labeled data for which the answers are already known, while unsupervised learning trains the model on data without label. Reinforcement learning trains the model to make decisions on the basis of feedback from the environment. Machine learning is applied in many industries, such as finance, agriculture, healthcare, transportation, retail, and more.

It has the ability to automate processes, increase accuracy, and provide insights and predictions [1–3].

- **Supervised Machine Learning**

Supervised machine learning is a type of machine learning in which a computer is trained on data that is labeled. You already know the correct answer because the computer has received the input data along with the correct output/target variables. The idea behind supervised machine learning lies in the training and testing of the model. The training phase involves learning the relationship between the data variables, and the testing phase involves building a model that is capable of making accurate predictions not only on the seen data but also on new data. The model is trained on a labeled dataset and then tested on another dataset that was not used in the training phase. The aim of the test is to determine the ability of the model to generalize to a new unseen dataset. Supervised machine learning can be used for a variety of purposes, such as image classification, speech recognition, and fraud detection. This type of machine learning is very useful when you have a large amount of data to work with [4].

- **Unsupervised Machine Learning**

In unsupervised machine learning, the algorithm learns from unstructured data to find patterns, structures, or relationships within the data. Unsupervised learning differs from supervised learning in that it does not have any labeled output values. Instead, the algorithm learns from the underlying structure or patterns of the data without any prior knowledge of what those patterns could be. When learning unsupervised algorithms, the goal is to find clusters, patterns, or relationships within the unstructured data based on its inherent properties. An example of an unsupervised learning technique is grouping similar data points together, while dissimilar points are grouped into various clusters. Another example of unsupervised learning is anomaly detection and association rule learning. Unsupervised learning is commonly used in experimental data analysis to get insights into the structure of the data, as well as in various applications such as: market segmentation, customer profiling, image and text analysis, and recommendation systems [5].

- **Reinforcement Learning**

In reinforcement learning, a computer program is trained to make decisions by testing and receiving feedback on how well or poorly the decisions were made. It is similar to training a puppy to perform a trick. When the puppy performs the trick correctly, it gets rewarded for doing so. In reinforcement learning, the computer program is the puppy and the reward is a treat. The computer program is put in a virtual environment and asked to complete a task. The computer program tries different tasks, and when it does well, it gets rewarded. Gradually, the computer program learns which actions result in the most rewards and begins to make better decisions over time. This form of learning can be used to teach computers to play games, to control robots, and even to make financial decisions. Reinforcement learning is a powerful tool for building intelligent systems that can learn and adjust to new situations [6].

In reinforcement learning, the agent does not receive explicit instructions about how to do a task. Instead, the agent learns from its interaction with the environment and makes decisions about what to do based on the reward for each action it takes. Reinforcement learning differs from supervised learning (which trains the agent on labeled data) and non-supervised learning (in which the agent learns from non-labeled data). There are many applications of reinforcement learning, such as playing games, controlling robots, and making financial decisions. Reinforcement learning is a powerful way to build intelligent systems that learn and adjust to new situations.

Supervised Machine Learning Models
In machine learning, classification and regression are two basic tasks that involve predicting the target variable based on the input features. Here is a brief description of classification and regression.

- **Classification:** A classification task is supervised in nature. The goal of classification is to predict a specific class or class for every input data point. In a classification task, the target variable is discrete and represents various classes or categories. Algorithms for classification learn from a labeled dataset. Each data point in the dataset is associated with a specific class label. The objective of classification is to construct a model that can correctly assign new unseen data points to a predefined class.

Examples of classification tasks are email spam detection, image classification, sentiment analysis, and regression. The goal of email spam detection is to categorize emails as "spam" or "not spam". Image classification is the process of identifying objects or animals within an image. For example, an image can be classified as a cat or a dog. Another supervised learning task is sentiment analysis. The goal of sentiment analysis is to identify the sentiment of text. The goal is to determine whether the text is positive, negative, or neutral.

- **Regression**: Like classification, regression is another supervised learning problem where the objective is to forecast a continuous numerical value (the target) based on the input features. In this case, the target variable is continuous, i.e., it can take any actual value within a given range. Regression algorithms also learn from labeled data. However, in regression, the labels are numerical.

Regression Algorithms

i. **Linear Regression**: It is one of the supervised machine learning algorithms used to forecast a continuous numerical output based on a single or multiple input variables. This algorithm finds the best linear relationship between input variables and output variable. The simplest form is simple linear regression, which involves only one input variable. In a 2D plot, the input variables are on the x-axis and the output variables are on the y-axis. The linear regression algorithm finds the best line for the data points, with m being the slope of the line and b being the Y-intercept, as shown in Eq. 3.1.

$$Y = mx + b \tag{3.1}$$

Linear regression can be used to forecast values like house price, stock price, and sales based on historical data, and it can be used to select features. You can evaluate the importance of each input parameter by looking at the regression coefficient. In these few assumptions are made about the underlying data, including linearity, homoscedastic (constant variances), and normalization of residuals, violations of these assumptions can affect model performance.

Linear regression can be used to predict stock prices, to estimate sales based on ad spending, to analyze the effect of variables on results, and much more. However, it should be noted that linear regression is based on the assumption that the relationship between variables is linear, which may not always be the case in real life. Linear regression can also be affected by outliers and can be restricted when dealing with complicated relationships among variables. In these cases, more sophisticated regression techniques or a nonlinear model may be more suitable [7, 8].

ii. Multiple Linear Regression

Multiple linear involves multiple independent variables. The relations between the variables are modeled as hyperplanes in a higher-dimensional space [9]. The multiple linear regression equation is an extension to the standard linear regression equation, as shown in Eq. 3.2.

$$y = b + m_0x_0 + m_1x_1 + m_2x_2 \ldots\ldots + m_nx_n \tag{3.2}$$

iii. Polynomial Regression

Polynomial regression, also known as logarithmic regression, is a type of nonlinear regression that can be used to model non-monotonic relationships between independent variables and dependent variables. Polynomial regression is different from linear regression; in that instead of fitting a linear line to the data, the polynomial curve is fitted to the data. Higher-order terms of independent variables are introduced into the regression equation, as shown in Eq. 3.3.

$$y = b + m_1x_1^1 + m_2x_2^2 + m_1x_3^3 \ldots\ldots. + m_1x_n^n \tag{3.3}$$

Here are some important points to keep in mind when using polynomial regression models:

- **Degree of Polynomial (n)**: The complexity of the curve fits the data better with a higher degree of polynomial. However, this can also lead to an overfitting of the data, especially if there are only a few data points.
- **Overfitting**: Polynomial regression can be used to detect nonlinear relationships. However, it is important to be careful about overfitting. A higher-degree polynomial can fit noise into the data and may not be able to generalize to non-visible data.

- **Feature Engineering**: With a polynomial regression system, you are essentially increasing the original features by raising them to different powers. As a result, some features may be highly correlated, which can be avoided by using ridge or lasso regression techniques.

Polynomial regression may be useful when you suspect that a relationship between variables is nonlinear.

iv. **Ridge Regression (L2 regularization)**: Ridge regression is a variant of linear regression that includes a regularization term in the loss function. It penalizes large coefficients to avoid overfitting. Ridge regression (commonly abbreviated to ridge or Tikhonov) is a type of linear regression used to solve the problems of multivariate and overfitting in models with many predictor variables. Ridge regression is a variant of regular linear regression in which an L2 periodization term is introduced into the cost function of the linear regression. The objective of regular linear regression is to find a coefficient that minimizes the sum of the squared difference between the actual and predicted values. Due to the high correlation between predictor variables, the estimated coefficients tend to become unstable in the face of small data changes. Ridge regression solves this problem by introducing a penalty term in the cost function, which is based on the Euclidean norm of a coefficient vector [10].

The benefits of ridge regression are that it can be used when there are multiple correlated predictor variables. This means that you do not have to completely exclude any predictor variables from your model if you believe that several predictors, even if correlated, are relevant. You can implement ridge regression using various Python programming libraries, such as scikit-learn. Choosing the appropriate λ (penalty) hyperparameter is critical and often requires fine-tuning techniques, such as cross-validations, to find the best value that does not overfit the data.

v. **Lasso Regression (L1 regularization):** Lasso regression is similar in nature to ridge regression but uses the regularization term L1 instead of L2. It not only penalizes overfitting but performs feature selection by reducing some feature coefficient to zero. Lasso regression, also known as least absolute shrinkage and selection operator (LASO), is a linear regression method that includes L1 regularization in linear regression cost function (LRA). Lasso regression is especially useful when dealing with large, high-dimensional datasets that contain many predictor variables. Lasso regression helps with feature selection by reducing some of the coefficients to absolute zero. The objective of standard linear regression is to minimize squared differences between predicted and observed values. However, if there are multiple predictor variables with some correlated values, the model may become unstable or overfitted the data [11].

vi. **Elastic Net Regression**: Elastic net combines L1 and L2 regression. Elastic net regression is a statistical approach to linear regression that combines the features of both L1 regularization and ridge regularization techniques. It is used to deal with situations where there are many predictor variables, and there may be high correlation between the predictor variables. In a regular linear regression model, the objective is to fit the model that predicts the target variable according to a collection of predictor variables. The goal is to find the coefficient that minimizes the sum of the squared difference between the predicted values and the actual values. When dealing with large numbers of features, the risk is that the model is overfitted (the model is too complicated) or that the model is unstable (the model has too many features), and the elastic net resolves these problems [12].

Classification Algorithms

i. **Logistic regression**

Logistic regression is a widely used for classification problem whose goal is to predict the class of observations by their properties. The logistic function assigns a real value from 0 to 1 to each input. This value can be interpreted as the probability of the dependent variable taking the value 1. It can be used to predict whether a customer will buy a product by looking at demographic and behavioral characteristics. It can also be extended to multiclass classification problems where the dependent variable takes more than two values, using a technique called "multinomial" logistic regression or "softmax regression" [13].

ii. **Support Vector Machine (SVM)**

SVM is the most popular supervised machine learning algorithm. It is commonly used for classification, regression, or outlier detection. SVM searches for a hyperplane that divides data points into different classes, with the largest difference between them being the hyperplane. The range is the distance between the hyperplane and the nearest data point in the class. The goal of SVM is to minimize the range while reducing the number of class errors. If data cannot be separated linearly, SVM maps it to higher-dimensional feature spaces using kernel functions. SVM can be used for multiclass classification by using either a single-sided or an all-sided approach. In a single-sided approach, a single binary classifier is trained for every pair of classes, and the results are combined. In an all-sided approach, each class is trained separately and the highest score is selected as a final prediction. SVM has several key advantages compared to other classification algorithms, such as can handle data with high- dimensionality, robustness to outliers, and ability to find unique solutions to specific problems. SVMs are sensitive to the choice of kernel function and its parameters, and can be computationally intensive on large datasets.

iii. **K-Nearest Neighbors (KNN)**

KNN is a straightforward and efficient machine learning algorithm used in both classification and regression tasks. KNN classifies or predicts a new data point by finding the closest neighbor to that point in the training dataset using a distance metric (e.g. Euclidean distance). The predicted class or value of that new point is determined by taking a majority vote of the nearest neighbors for classification or averaging their values for regression. The K parameter is an important parameter for KNN because it affects the tradeoff between bias and variance. A small K parameter produces less distortion and more variance, while a larger K parameter produces more distortion and less variance. The K parameter can be selected by cross-validation and other methods. KNN has a few advantages like simplicity, flexibility, and non-linear relationships between features and target. KNNs are especially useful when the decision boundary between classes or the relationship between features and the target variables is nonlinear and complex. KNNs are also commonly used in recommender and anomaly detection applications.

iv. **Decision Tree**

Decision trees are one of the most common machine learning algorithms used for classification and regression tasks. Decisions trees work by constructing a tree-style model of the decisions and their implications. Decision trees begin with a single tree node representing the entire dataset. The algorithm then selects the most suitable feature to divide the data according to certain criteria like entropy or Gini index. The features with the highest information gain or lowest contamination are chosen to divide the data into 2 or more sub-datasets. The process is repeated for each sub-dataset until a stopping point is met (e.g., maximum tree depth, minimum number of samples per leaf nodes). Decision trees can be visualized like flowcharts, which makes it easier to understand and explain. The decision tree is used to predict new data points. Each inner node represents a decision according to a characteristic, and each outer node represents a class/value.

Decision trees outperform other ML algorithms in several ways, such as:

- Handling both categorical functions and continuous functions.
- Interpreting nonlinear relationships.

However, decision trees are prone to overfitting, particularly if the tree has too many features or is too deep. To avoid this, various techniques can be employed, such as pruning, maximum tree depth setting, or ensemble techniques like random forests or gradient boosting.

Ensemble Learning

It is the process of combining the predictions of several different models (learners), in order to create a more powerful and precise predictive model. The fundamental idea behind this approach is that the strengths of each model can compensate for the weaknesses of the others, resulting in improved overall performance. Ensemble methods work best when dealing with complicated and difficult problems [14]. They often result in better generalization and more robustness. There are a number of popular ensemble techniques discussed in the following section.

i. Random Forest

Random forest is the most widely used ensemble learning algorithm. It is used for both regression and classification tasks. The idea behind it is to create several decision trees and combine their predictions to minimize overfitting and enhance model accuracy. Each decision tree in a random forest trains on a random set of training data and a random set of features. The randomness helps to minimize correlations between trees and increases diversity. The predictions from each tree can be combined using aggregation techniques such as majority voting for classification or mean for regression. Random forests offer several advantages over one decision tree:

- High-dimensional data handling.
- Roughness against outliers or noisy data.
- Ability to capture nonlinearly related features.
- Efficiency for large datasets.

Fine-tuning hyper parameters can improve random forest performance. Random forests are used in a wide range of applications, including classification, regression, feature selection, and anomaly detection. Random forests are especially useful when the decision boundary between classes or the relationship between features and target variables is complicated and nonlinear.

ii. Gradient Boosting

Gradient boosting is one of the ensemble methods that builds a robust model by combining predictions of several weak models, usually decision trees. Gradient boosting builds trees in a sequential manner, with each tree correcting the previous one's errors. The algorithm begins with a first prediction, usually a simple estimate, such as the mean (for a regression) or the constant (for a classification). Each boosting round involves training a weak learner to capture the errors (or residuals) of the ensemble's previous predictions. Gradient boosting uses gradient descent optimization to reduce a specific loss function, iteratively adjusting the predictions to minimize the loss. A key element is the implementation of a learning rate that scales the impact of every new learner, balancing model complexity with accuracy. The predictions of all learners are weighed and combined to form the final prediction. This iterative approach continues until a predetermined stopping criterion is met (i.e., number of boosting rounds or desired level of accuracy). Gradient boosting's strengths are its ability to deal with complicated relationships in data, automatic feature selection, and producing highly accurate models (although careful hyperparameter fine-tuning is required to unlock its full potential).

iii. Extreme Gradient Boosting (XGBoost)

Extreme gradient boosting is a powerful machine learning algorithm that combines the predictions of several decision trees to build a powerful predictive model. It is known for its high performance and versatility when it comes to predictive modeling tasks. XGBoost is one of the most popular machine learning algorithms developed by

Tianqi Chen. It has gained a lot of popularity among the machine learning community and is often used as the default algorithm for various applications, such as regression and classification, ranking and recommendation systems. One of the key differences between XGBoost and other algorithms is the optimization techniques and the regularized learning framework that it uses. Gradient boosting is the iterative improvement of a model by minimizing a specific loss function. In XGBoost, you can combine the predictions of different decision trees to create a powerful model. XGBoost's key features include the ability to work with both large and small amounts of data, the ability to handle missing values automatically, and the ability to work in parallel, making it more efficient for large datasets. XGBoost also offers various hyperparameter tuning options, which allow practitioners to fine-tune their models for the best possible results. Predictive accuracy is one of the main reasons for its popularity. Built-in feature importance scores and visualization tools also help in understanding and interpreting the model's behavior, making it more appealing in academia and industry. Concisely, we can say that XGBoost has earned its place as one of the most trusted and powerful tools in the machine learning stack, appreciated for its ability to deal with complex data, provide high-quality results, and enable meaningful model insights. Whether you are using XGBoost to solve a classification problem, a regression problem, or a ranking problem, you can be sure that it is one of the best tools in the toolbox for data science and machine learning.

Unsupervised Machine Learning Models

i. **K-means Clustering**

K-means is one of the most commonly used cluster algorithms in machine learning (ML) and data analysis. The main purpose of K-means is to divide a dataset into different, non-interrelated groups or clusters based on how similar the data points are. The algorithm works in an iterative manner and is relatively simple. First, K-means starts by randomly choosing a set number of clusters, often called "K" centroids. These initial K centroids act as the initial points of entry into the clusters. After that, the algorithm iterates through each iteration as follows:

– K-means assigns each data point to the cluster that has the nearest centroid for each of the data points in the dataset. This step is repeated for all data points. This reassigns each point to the closest cluster.
– After all data points are assigned to the clusters, the algorithm calculates the new cluster centroid by taking the average of all the data points in each cluster. The new centroid represents the center of the cluster.
– K-means performs a convergence check until the centroids are converged. Convergence occurs when the centroid does not change significantly between iterations or when a certain number of iterations are reached.

The goal of K-means is to find centroids that minimize the sum of the squared distances between the data points and their cluster centroids i.e. Within-Cluster Sum of Square (WCSS). WCSS is a measure of the compactness of a cluster. The K-means algorithm has several benefits, such as simplicity, efficiency, and scalability. It can be used to cluster large datasets. However, K-means has some drawbacks. It is sensitive to the initial location of the centroids, requires the number of clusters to be specified in advance, and assumes that clusters are approximately spherical and equal in size.

ii. **DBSCAN**

DBSCAN stands for Density-Based Spatial Cluster of Applications with Noise. DBSCAN is a noise-based unsupervised learning algorithm used to group data points according to density distribution. DBSCAN differs from traditional K-means algorithms in that it does not assume that clusters are spherical or have a predetermined number of clusters. This makes DBSCAN especially useful for finding clusters of any shape and for dealing with noisy data, working of DBSCAN is as follows:

– Core points and density: A core point in DBSCAN is a data point with at least a certain number of neighboring data points (epsilon or ε) within a defined distance (minPts). Core points are the central points of the cluster. The minPts parameter and the ε parameter are user defined and are important for the behavior of the algorithm.
– Border points: A border point is a data point within ε distance from a core point but does not have enough adjacent data points to be considered a core point. Border points are located on the edge of the cluster and are a part of the cluster, but not central to the cluster.
– Noise points and outliers: Data points that are not a core point or border point are considered "noise points" or outliers. They do not belong to any cluster and can be noise or irrelevant data.
– Clustering: The DBSCAN cluster process begins by selecting an unvisited, random data point. DBSCAN then checks if the data point is a key point. If the key point is present, DBSCAN expands the cluster by repeating the process until all reachable key points (and their boundary points) are in the same cluster. DBSCAN continues to expand the cluster until no more key points can be added. DBSCAN then selects another random data point and continues the process, expanding the cluster as required.
– Termination: The cluster process continues until all visited data points are visited, assigned to clusters, or are labeled as noise points.

The main strengths of DBSCAN are its ability to identify clusters of different shapes, its ability to cope with noise and outliers, and its freedom to determine the number of clusters predetermined. However, selecting the correct values for ε, and minPts, can be difficult and may necessitate domain knowledge or trial and error. DBSCAN can also struggle with clusters of very different density. The DBSCAN algorithm is widely used in the following fields: spatial data analysis, image segmentation, anomaly detection, and customer segmentation. It is a versatile and powerful algorithm for finding clusters in datasets where there is no well-defined distribution of clusters or where the number of clusters.

3.3 Deep Learning Models

Deep learning algorithms are highly sought after because of their ability to meet several fundamental requirements in the area of AI and ML. One of the fundamental requirements that deep learning meets is the ability to solve complex pattern recognition problems. In applications such as image and speech recognition (ICT), Natural Language Processing (NLP), and Computer Vision (CV), deep learning excels at automatically learning hierarchical representations (HDRs) of data. HDRs allow deep learning algorithms to capture complex features and relationships in the data, in which traditional machine learning methods are unable to do. Another important requirement that deep learning meets is the capacity to process large amounts of data. In today's data-centric world, where organizations produce huge amounts of data, the ability of deep learning algorithms to process and extract valuable information from large datasets makes them essential for tasks such as big data analytics where traditional approaches often struggle with scalability and performance.

In addition, deep learning algorithms are great at feature learning. Deep learning algorithms can automatically find and extract features from raw data without the need for a lot of manual feature engineering. Not only does this save time, but it also improves performance, especially in situations where you have high-dimensional data, and it can be hard to find the right features.

Another critical need is demonstrated by deep learning's impressive performance in many machine learning benchmarking and competitions. From image and object classification and speech recognition to natural language understanding and more, deep learning is constantly pushing the limits of what is possible. With its high accuracy and capabilities, deep learning is the go-to choice for tasks that demand high performance. Neural networks are also known for their flexibility and generalization. Neural networks can be used for a variety of purposes, such as image processing, speech recognition, language processing, recommendation processing, and even autonomous systems. With this flexibility, organizations can use deep learning across a variety of areas and applications, yielding useful insights and solutions [15].

i. **Convolutional Neural Networks (CNNs)**

Conventional neural networks (CNNs) are deep learning architectures specifically designed for processing and analysis of visual data. CNNs have revolutionized computer vision and image processing. Inspired by the human visual cortex, CNNs can be used to classify images, identify objects, recognize faces, and perform more complex tasks such as image segmentation, style transfer, and more. One of the key innovations of a CNN is its convolutional layers. CNNs use convolutional filters, or kernels, to transform input images into hierarchical features. These filters start from simple edges and textures and work their way up to more complicated shapes and patterns. CNNs often employ pooling layers to downsize feature maps and minimize computational complexity while preserving important information. CNNs are widely used in self-driving vehicles, medical image processing, and video analysis, as well as in natural language processing (NLP) tasks such as text classification, sentiment analysis, and word embedding, where text data is converted into an image-like representation. CNNs have greatly advanced our capacity to process and comprehend visual information and have become the foundation of modern artificial intelligence (AI) and machine learning.

ii. **Recurrent Neural Networks (RNNs)**

Recurrent neural networks or RNNs are artificial neural networks that are designed to process sequential data. RNNs are different from feedforward networks because they have a built-in memory or context that persists over time. This makes RNNs extremely useful for time series data processing, NLP, speech recognition, etc. What makes RNNs so useful is that they have recurrent connections. A recurrent connection is a loop in a neural network that allows information to pass from one step of a sequence to another. RNNs can capture dependencies or patterns in sequence data by taking into account previous information in relation to the current input. Standard RNNs, however, are limited in capturing long-distance dependencies because they suffer from a vanishing gradient problem. This problem occurs when gradients shrink too much during training, making it difficult to learn long-distance dependencies. To overcome this issue, more sophisticated RNN variants have been developed, such as LSTM and GRU. These variants include specialized gate mechanisms that help to reduce the vanishing gradient problems and allow for the modeling of long-lived dependencies. RNNs are widely used in NLP, machine translation, sentiment analysis, speech recognition, and time series prediction. They are also used in generative models to generate sequences of data. RNNs' ability to process sequential information makes them an essential tool in deep learning and AI, allowing for modeling and understanding of intricate temporal relationships in data.

iii. **Long Short-term Memory (LSTM)**

LSTM stands for "long short-term memory". Long-term memory is a term used to describe a type of architecture of recurrent neural networks (RNNs) that is designed to handle long sequences and capture long-term relationships in data. LSTMs was first introduced in 1997 by the team led by Matthias Schroder. Since then, LSTMs have become an essential part of many applications that involve sequential data. Examples of applications that use LSTMs include NLP, speech recognition (speech synthesis), machine translation, and time series forecasting. LSTMs have gained popularity because they are able to learn and recall information over long periods of time, making them great at working with sequential data with intricate patterns and relationships.

LSTMs are characterized by their architecture, which consists of three key elements:

Memory Cell: The LSTM contains a memory cell that stores information over several time steps. The memory cell is self-regulating and controls what information is written into, read out of, or erased from the memory cell, allowing the LSTM to retain important information while discarding irrelevant information.

Forget Gate: LSTM uses a forget gate to decide which information from the preceding time step is to be forgotten or retained within the memory cell. The forget gate takes the previous hidden state as input and the current input as output to activate the forget gate, which scales the memory cell values.

Input and Output Gates: The LSTMs have inputs and outputs gates that control the flow of information from the memory cell into the LSTM. These gates determine which new information is to be written into the memory cell and which information is to be exposed as output to the LSTM during the current time step.

LSTMs combine these gates with the memory cell to capture long-lived dependencies in data, reduce the vanishing gradient problem in traditional RNNs, and make them suitable for tasks that involve the understanding and generation of sequences of data. LSTMs have been instrumental in advancing the state of the art in various domains, such as NLP, speech recognition, and autonomous systems such as self-driving cars where they are utilized to process and interpret sequence sensor data. Their ability to model and remember intricate temporal patterns has made LSTMs a key component of modern deep learning and AI.

iv. **Generative Adversarial Networks (GANs)**

Generative neural networks (GNNs) are artificial neural networks designed to create new data samples that are similar to the dataset on which it has been trained. GNNs are a subset of deep learning models used in machine learning. They are especially useful for content creation tasks, such as image, text, music, or even video. The generator and the discriminator are the two main parts of a general-purpose neural network. The generator generates data samples, and the discriminator verifies whether the generated data is real or not. The generator learns to generate data that is more and more indistinguishable from the real data. The discriminator learns to distinguish between the real data and the fake data. This is known as adversarial training, or GAN, which improves the generator's ability to generate realistic data. GANs can be used in a wide range of applications, from image synthesis to text generation. They can even be used to discover drugs. Generative models can be used to solve specific problems and produce high-quality output. Generative models have been used to solve a variety of problems, such as variable auto encoders or transformers. Generative networks have demonstrated remarkable creativity and are able to generate new and realistic data.

3.4 Large Language Models

Large language models (LLMs) are the future of AI and NLP because of their transformative power. They are the answer to the growing need to understand, generate, and interact with human speech. LLMs are used to enhance natural language understanding and text generation, as well as to support multilingual communications.

LLMs have the advantage of transfer learning, which accelerates the development and implementation of NLP based on pretrained knowledge. They have human-like semantic understanding and text generation capabilities in content creation, science research, business, and accessibility. But as these models become integrated into different areas, ethical considerations and responsible use of AI continue to be paramount. The demand for LLMs reflects their potential to transform the way we communicate with AI systems and open up new opportunities in communication, access to information, and content generation.

LLMs are capable of understanding and processing natural language text at a high level of accuracy and complexity. This is important for applications that need to understand the nuances of language, including sentiment analysis and text classification, efficient text generation, entity recognition, etc. These models are good at generating text that is consistent and relevant to the context. They are useful for applications such as: content generation, chatbots automated content summaries, creative writing, and critical text quality.

Multilingual Support: LLMs can support multiple languages, which makes them important for tasks such as machine translation, multilingual information retrieval, multilingual customer support, and more.

Transfer Learning: Pre-trained LLMs are a great starting point for various NLP tasks. These models can be fine-tuned for specific tasks, requiring less labeled data and computing power compared to training from the ground up. This means that NLP solutions can be developed and deployed more quickly.

LLMs are capable of semantic understanding, which allows for more sophisticated natural language processing and reasoning. This is useful for tasks such as question and answer, content recommendation, and search engine optimization. Some LLMs specialize in human-like text generation. These models are designed to produce text that closely mimics human-generated content. This makes them useful for marketing, journalism, and creative writing.

Scientific research large language models can be used in scientific research to help review literature, generate hypotheses, and interpret data across a wide range of domains. These models can improve accessibility for people with disabilities by providing speech-to-speech (CTS) capabilities, facilitating communication, and improving accessibility of content. LLMs can be used in a variety of business applications, ranging from improving customer support through chatbots to improving user experiences via personalized recommendations and content.

LLMs play an essential role in facilitating more natural and interactive human-to-machine interactions with AI systems, including virtual assistants, voice-controlled devices, etc. LLMs can help in content moderation in online platforms by detecting and flagging content that is inappropriate or harmful. The development and implementation of LLMs has raised important ethical concerns, such as: bias detection and mitigation, fairness in AI, privacy protection, and responsible AI deployment.

LLMs are characterized by their extremely large number of parameters. This large number of parameters enables them to capture complex relationships and patterns in the language. These models are pretrained on large text corpora for general language comprehension. After pretraining, these models can be fine-tuned for specific tasks, allowing them to be adapted to various NLP applications. The majority of LLMs use the transformer architecture, which uses self-attentive mechanisms to efficiently process and generate sequence of text data.

LLMs are good at NLP, and they can be used to classify text, analyze sentiment, and recognize named entities. LLMs can also be used for text generation, and they can generate text that is consistent and contextually relevant. LLMs can be used to create content, build chatbots, and write creative content. Large language models are also useful for machine translation. They can help improve machine translation systems by providing high-quality translation across multiple languages.

They are well-suited for question and answering (Q&A) where they can answer questions in natural language, making it easier to build chatbots and virtual assistants. Some large models are multimodal, meaning they can understand and generate content from text as well as other modalities such as images and audio. Transfer learning large language models can be used as a base model for many NLP tasks, eliminating the need for task specific training data and accelerating the development process of NLP apps.

Some of the most well-known large language models are OpenAI's GPT (generative pretrained transformer) series and Google's BERT (bidirectional encoder representations from transformers), as well as other models developed by various organizations and researchers around the world. Large language models have had a significant impact on the advancement of AI-powered natural language processing (NLP) and language generation and continue to play an important role in a variety of NLP apps and research efforts.

3.5 Summary

Machine learning, deep learning, and large language algorithms are the key building blocks of modern AI. Each has its own unique strengths and advantages in the field. Machine learning focuses on pattern detection and predictive modeling. It has revolutionized the way data-driven decisions are made across industries. With its ability to learn from large datasets, automate processes, and improve problem-solving, machine learning has become an essential tool in today's data-driven world.

Deep learning is a subset of machine learning. Deep learning has become a dominant technology in tasks that involve large amounts of data, including images, text, and sequence data. Deep neural networks, which are the brains behind deep learning models, have led to advances in computer vision and natural language processing as well as speech recognition. Deep learning models have a hierarchical feature learning structure that enables them to understand complex relationships and patterns in data. Larger language algorithms (a subset of deep learning) have revolutionized the way we understand and generate natural language. With their vast scale and pretrained understanding, these models can understand, generate, and engage with human speech at a level we have never seen before. They have found use cases in content creation, translation, question and answer, and more, changing the way we interact with AI systems. All these algorithms illustrate the development of AI, ranging from machine learning, deep learning, and large-scale language models. These algorithms address the increasing complexity of challenges, the proliferation of information, and the demand for sophisticated natural language comprehension. However, they must be developed and deployed in a way that is ethical, privacy conscious, and biased in order to maintain AI's benefits for society while reducing risks. As these areas continue to develop, their combined effect on technology and business, as well as on society, is likely to become even more significant in the coming years.

References

1. Zhou ZH (2021) Machine learning. Springer Nature
2. Mitchell TM (1997) Machine learning
3. Alpaydin E (2021) Machine learning. MIT Press
4. Nasteski V (2017) An overview of the supervised machine learning methods. Horizons b 4:51–62
5. Hahne F, Huber W, Gentleman R, Falcon S, Gentleman R, Carey VJ (2008) Unsupervised machine learning. Bioconductor Case Stud 137–157
6. Kaelbling LP, Littman ML, Moore AW (1996) Reinforcement learning: a survey. J Artific Intell Res 4:237–285
7. Su X, Yan X, Tsai CL (2012) Linear regression. Wiley Interdisc Rev: Comput Stat 4(3):275–294
8. Aiken LS, West SG, Pitts SC (2003) Multiple linear regression. Handbook Psychol 481–507
9. Regression. R Through excel: a spreadsheet interface for statistics, data analysis, and graphics 269–284.
10. McDonald GC (2009) Ridge regression. Wiley Interdisc Rev: Comput Stat 1(1):93–100
11. Ranstam J, Cook JA (2018) LASSO regression. J British Surg 105(10):1348–1348
12. Xu QF, Ding XH, Jiang CX, Yu KM, Shi L (2021) An elastic-net penalized expectile regression with applications. J Appl Stat 48(12):2205–2230
13. Das A (2021) Logistic regression. Encyclopedia of quality of life and well-being research. Springer International Publishing, Cham, pp 1–2
14. Zhou ZH, Zhou ZH (2021) Ensemble learning. Springer, Singapore, pp 181–210
15. Goodfellow I, Bengio Y, Courville A (2016) Deep learning. MIT press

Chapter 4
Machine Thinking: New Paradigm Shift

4.1 Machine Learning to Machine Thinking

Taking appropriate decisions considering human psychology by machine is a need of today. The current AI models are data intensive and require a huge amount of data for model building. There are many business cases in computing which merely require not only machines to learn but machines also need to think. As discussed in earlier chapters of this book, there are many critical applications, where existing data-centric AI is inadequate and there is a need for machine thinking-based applications. It is very important to understand how machine learning has been transforming over the decades and its current state [1, 2].

Machine learning has undergone significant evolution since its inception, both in terms of algorithms and applications. Here are some key stages of its evolution:

- **1950s-1960s**: The foundations of machine learning were laid during this period. Researchers focused on developing techniques such as linear regression, decision trees, and Bayesian networks.
- **1970s–1980s**: Machine learning faced limitations due to limited computational power and data availability. Symbolic learning and expert systems dominated this era, with a focus on rule-based reasoning.
- **1990s–2000s**: Machine learning witnessed advancements in neural networks and statistical learning. Support Vector Machines (SVM), Hidden Markov Models (HMM), and ensemble methods gained popularity. Feature engineering played a significant role during this time.
- **2010s**: Deep learning emerged as a game-changer in machine learning, enabled by advancements in computational power and availability of large datasets. Deep neural networks with multiple layers revolutionized tasks such as image recognition, natural language processing, and speech recognition.

M. R. Velankar et al., *Cognitive Computing for Machine Thinking*, Innovations in Sustainable Technologies and Computing, https://doi.org/10.1007/978-981-97-0452-1_4

- **Present Day**: Machine learning is increasingly fueled by big data and cloud computing. There is a growing emphasis on unsupervised learning, reinforcement learning, and interpretable models. Techniques like transfer learning and generative adversarial networks (GANs) have also gained attention.

This evolution of machine learning clearly indicates that with the huge increase in data as well as the requirement of high-performance computing applications, nowadays machine learning and all its subsets' technologies are also becoming inadequate [3, 4]. Machine learning models heavily rely on large amounts of high-quality data for training. Limited or biased data can lead to inaccurate or biased models. Machine learning models may over fit the training data, resulting in poor generalization to unseen data. Overfitting occurs when the model becomes too complex and memorizes noise in the training data instead of learning underlying patterns. Black-box nature of some machine learning models, such as deep neural networks, can make it challenging to understand and interpret their decisions. Explainable AI techniques aim to address this limitation. Machine learning models are sensitive to the quality and preprocessing of the input data. Outliers, missing values, and inconsistent data can impact the model's performance and reliability. Machine learning models often require domain expertise to design relevant features and select appropriate representations. Feature engineering can be time-consuming and may require expert knowledge.

In addition to this, complex machine learning models, especially deep learning models, require significant computational resources, including processing power and memory. Training and deploying such models can be computationally expensive. Machine learning models can perpetuate biases present in the training data, leading to biased outcomes. Unconscious bias in data collection, human annotation, or model design can impact fairness and equity. Machine learning models may struggle to adapt to changes in the distribution of the data they were trained on. Concept drift, where the relationship between input and output changes over time, poses a challenge to maintaining model performance. Machine learning models, especially those trained on sensitive data, raise concerns about privacy and potential data breaches. Models can inadvertently leak private information embedded in the training data. Some machine learning models require significant computational resources and time for training and deployment, limiting their scalability and real-time applications.

In the sequel, machine thinking [5–7] is a need of today. Machine thinking refers to the set of methodologies and culture used by humans to teach machines how to advance toward a design goal. It involves designing algorithms and models that enable machines to learn from data and make decisions or predictions based on that learning. Machine thinking refers to different aspects of artificial intelligence, such as designing algorithms, models, and architectures that enable machines to learn, reason, and make decisions. The motivation behind machine thinking can vary depending on the application and context, ranging from improving efficiency, accuracy, or productivity to creating new opportunities and solving complex problems where machines need to think. Machine thinking is a key aspect of artificial intelligence and is used in a wide range of applications, from self-driving cars to natural language processing. Although, "machine thinking" is not a widely accepted term in artificial intelligence, there is no agreed-upon set of objectives that can be attributed to it. However, the objectives of artificial intelligence, which is a broader field that encompasses machine learning and other related subfields, can shed some light on the potential objectives of "machine thinking". One of the main objectives of machine thinking is to automate tasks that are currently performed by humans, such as image recognition, natural language processing, and decision-making. Machine thinking can also be used to enhance human capabilities in various domains, such as healthcare, education, and finance. For example, it can help doctors diagnose diseases more accurately or enable personalized learning for students. Machine thinking can also help organizations and businesses streamline their operations and increase efficiency and productivity. For example, it can be used in logistics to optimize routes and reduce delivery times. Machine thinking can enable the creation of new products and services that were not possible before. For example, thinking-powered chatbots can provide customer service 24/7, or AI-generated music can provide new forms of entertainment. To summarize, machine thinking mainly enables automating tasks, enhancing human capabilities, or solving complex problems.

4.2 Human and Computer Decision-making

Human and computer decision-making refers to the process by which decisions are made in situations where both humans and computers are involved. This can include situations where humans rely on computers to provide information or make recommendations, situations where humans and computers work together to make decisions, or situations where computers make decisions autonomously based on predefined rules or machine learning algorithms. The interaction between humans and computers can have an impact on the final decision that is made, and research in this area aims to explore how to best leverage the strengths of both humans and computers to make effective decisions. This can involve designing interfaces and decision-making processes that are intuitive and easy for humans to use, or developing algorithms and machine learning models that can effectively process and analyze large amounts of data. Human and computer decision-making can work in different ways depending on the specific context and application. Some general ways in which human and computer decision-making can work are listed below:

1. **Human-in-the-Loop**: In this approach, computers provide information or recommendations to humans, who then make the final decision based on their own judgment. For example, in medical diagnosis, a computer program can analyze patient data and provide a list of possible diagnoses, but a doctor ultimately makes the final diagnosis based on their own expertise and judgment.

2. **Computer-in-the-Loop**: In this approach, humans provide input to a computer program, which then makes the final decision based on predefined rules or machine learning models. For example, in credit scoring, a person's creditworthiness can be assessed based on a computer program that analyzes their credit history and other factors.
3. **Collaborative Decision-making**: In this approach, humans and computers work together to decide based on their respective strengths and expertise. For example, in stock trading, a human trader can use a computer program to analyze market trends and make trading decisions based on their own judgment and the program's recommendations.
4. **Autonomous Decision-making**: In this approach, computers make decisions autonomously based on predefined rules or machine learning models without human intervention. For example, in self-driving cars, the car's computer system makes decisions about steering, acceleration, and braking based on input from sensors and machine learning models.

The goal of human and computer decision-making is to leverage the strengths of both humans and computers to make effective and efficient decisions. The specific approach used will depend on the application and the context, as well as the availability of data and expertise. There are several technologies used in human and computer decision-making. Clinical decision support systems are computer programs that provide health care professionals with information and tools to support clinical decision-making. They can include tools for diagnosis, treatment, and monitoring. AI technologies, such as machine learning and natural language processing, can be used to analyze large amounts of data and provide insights to support decision-making. Computerized physician order entry is a technology that allows health care professionals to enter medication orders and other clinical orders into a computer system. This can help to reduce errors and improve patient safety. Automated decision-making involves the use of data, machines, and algorithms to make decisions in a range of contexts, including public administration and finance. Internet of Things technologies can be used to collect data from sensors and devices, which can then be analyzed to support decision-making. For example, IoT sensors can be used to monitor energy use in buildings and optimize energy efficiency.

Human and computer decision-making can work in different ways, depending on the specific context and application. Computers can provide information and recommendations to humans, or humans can provide input to computer programs. In some cases, humans and computers work together collaboratively to make decisions, while in other cases, computers make decisions autonomously. Clinical decision support systems are computer programs that provide information and tools to support clinical decision-making. They can be used to augment clinicians in their complex decision-making processes. These systems can consider all data available in the electronic health record, making it easier for clinicians to make informed decisions. AI technologies, such as machine learning and natural language processing, can be used to analyze large amounts of data and provide insights to support decision-making. AI can boost our analytic and decision-making abilities and heighten creativity. Both humans and computers can contribute to decision-making in different ways. The specific approach used will depend on the application and the context, as well as the availability of data and expertise.

4.3 Machine Thinking Models

Alen Turing in 1950 raised a very interesting concern about the lack of a machine thinking model and coined a very strong statement.

> I believe that at the end of the century the use of words and general educated opinion will have altered so much that one will be able to speak of machine thinking without expecting to be contradicted.
>
> - Alan Turing 1950

However, it should be noted that the existing studies on machine thinking are without any definitions and concrete references. The main reason is that understanding the brain is the most challenging task. The brain understanding requires the convergence of biology, neuroscience, mathematics, computer science, psychology, and electrical engineering and their application to the underlined applications. Ideally, machine thinking requires key operations which includes attention, sensing, action, and context. Machine thinking can be realized with the Universal Turing Machines (UTM) which represent a look-up table with a symbolic network of finite automata. Machine thinking is like animal-like thinking which is general, learned via sensorimotor, which has covert parts and displayable parts and is fully autonomous [8]. In this analogy the thinking with abstraction is depicted in Fig. 4.1.

1. **Conscious Modeling**

The key objective of conscious models is artificial consciousness or the potential for machines to become self-aware or conscious. This is a highly debated topic in the field of AI and there are different opinions on whether it is possible for machines to achieve consciousness. Some researchers argue that consciousness is not just a matter of processing power, but also involves subjective experience and a sense of self. Others believe that it may be possible to create conscious machines using advanced neural networks, cognitive architectures, or other AI techniques. However,

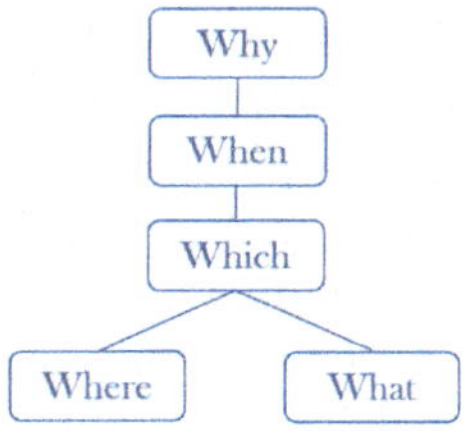

Fig. 4.1 Thinking with abstraction

the idea of a conscious model in AI is still speculative and remains an active area of research and debate. This model includes four stages:

(a) **Unconscious Incompetence**

Unconscious incompetence is a term used in the "Four Stages of Competence" model to describe a stage of learning in which an individual is not aware that they lack knowledge or understanding in a particular area. At this stage, the individual does not understand or know how to do something and may not even recognize the deficit. This is the first stage in the learning process, and it is characterized by ignorance or lack of awareness. The term can be applied in various contexts, including education, training, and personal development. The Four Stages of Competence model is often used to help learners understand their own learning process and develop new skills more effectively.

(b) **Conscious Incompetence**

At this stage, an individual is aware of a skill or knowledge gap and recognizes the importance of acquiring the new skill. In this stage, the individual understands that they lack knowledge or understanding in a particular area and may feel uncomfortable or frustrated because of this awareness. This stage is characterized by a willingness to learn and a desire to improve. The term can be applied in various contexts, including education, training, and personal development. This model is often used to help learners understand their own learning process and develop new skills more effectively.

(c) **Conscious Competence**

At this stage, the individual acquires the necessary knowledge or skill, but it requires conscious effort and practice to execute it effectively.

(d) **Unconscious Competence**

At this stage, the individual has acquired the knowledge or skill to the point where it can be executed effortlessly and without conscious thought.

2. **Cognitive Modeling**

Cognitive modeling [9–11] is a field of computer science and psychology that involves creating computational models of human cognitive processes. These models are designed to simulate and understand how people think, learn, reason, and solve problems. Cognitive modeling is used in a variety of applications, including artificial intelligence, human–computer interaction, psychology research, and machine thinking. By creating computational models of cognitive processes, researchers can gain insights into how the human mind works and develop better ways to design technology and support human learning and decision-making. Computational models of cognition can be used to test theories and hypotheses, make predictions, and develop new interventions and treatments for cognitive disorders. Stages of cognitive modeling are listed in Fig. 4.2.

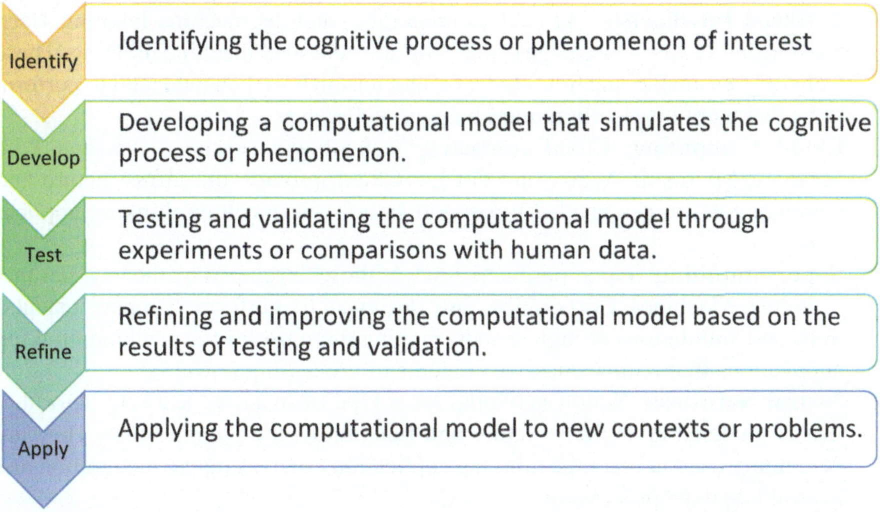

Fig. 4.2 Stages of cognitive modeling

These steps may be iterative, with the model being refined and improved over time based on new data and insights. However, it's important to note that these steps may vary depending on the specific application and context of cognitive modeling.

There are several algorithms used in cognitive modeling, particularly in the development and refinement of computational models that simulate cognitive processes. Some examples of algorithms used in cognitive modeling include:

- **Connectionist Algorithms**: These algorithms are used in neural network models, which simulate the way neurons in the brain process information and communicate with each other.
- **Rule-based Algorithms**: These algorithms are used in production system models, which simulate the way humans use rules and heuristics to solve problems and make decisions.
- **Instance-based Algorithms**: These algorithms are used in case-based reasoning models, which simulate the way humans retrieve and apply knowledge from past experiences to solve new problems.
- **Reinforcement Learning Algorithms**: These algorithms are used in models that simulate the way humans learn from feedback and rewards and adjust their behavior accordingly.

4.4 Application Development

Application development in machine thinking is the new paradigm shift and relevant technologies are listed below:

1. **Artificial Intelligence (AI)**: AI technologies, such as machine learning, deep learning, and natural language processing, are central to machine thinking. These technologies enable machines to learn, and reason based on data, and to perform tasks that would typically require human intelligence.
2. **Cloud Computing**: Cloud computing technologies, such as Amazon Web Services, Microsoft Azure, and Google Cloud, provide the infrastructure and computing resources needed to support large-scale machine thinking applications.
3. **Supercomputing**: Supercomputers, such as those developed by the now-defunct Thinking Machines Corporation, are designed to perform complex calculations and simulations at high speeds, making them useful for machine thinking applications that require massive amounts of processing power.
4. **Neural Networks**: Neural networks are a type of machine learning algorithm that is modeled after the structure and function of the human brain. They are commonly used in machine thinking applications such as image recognition and natural language processing.
5. **Von Neumann Architecture**: The von Neumann architecture is a computing model that is based on the idea of a stored-program computer, where data and instructions are stored in the same memory system. This architecture is commonly used in machine thinking applications that involve complex computations and decision-making processes.

Cognitive computing and machine thinking platforms are technology platforms that utilize artificial intelligence and signal processing techniques to enable various functionalities such as machine learning, reasoning, natural language processing, speech recognition, vision (object recognition), human–computer interaction, dialog, and narrative generation, among others. These platforms are designed to process and analyze large, unstructured datasets and provide intelligent insights and solutions. IBM Watson is a well-known cognitive computing platform that offers a range of services and APIs for developers to build intelligent applications. It leverages machine learning, natural language processing, and visual recognition technologies to provide advanced cognitive capabilities. Microsoft Azure provides a suite of cognitive services that enable developers to easily integrate artificial intelligence capabilities into their applications. These services include language understanding, speech recognition, computer vision, and more.

Google Cloud AI offers a variety of cognitive computing services, including natural language processing, speech-to-text, text-to-speech, and image recognition. These services can be used to build intelligent applications and enhance user experiences. Amazon web and AI services provides a range of AI services, including text and speech analysis, image and video analysis, and natural language processing. These services can be used to build cognitive applications that can understand and interpret human language and behavior. These platforms provide developers with the tools and resources necessary to build cognitive applications that can understand, reason, and interact with humans in a natural manner. They offer APIs and SDKs that simplify the development process and enable the integration of cognitive capabilities into existing applications.

It's important to note that cognitive computing platforms are constantly evolving, and new platforms may emerge in the future. Developers should consider their specific requirements and choose the platform that best suits their needs.

Developing machine thinking applications typically involves several steps, which are listed below:

- **Define the Problem**: Identify a specific problem that can be addressed using machine thinking. This could be anything from image recognition to natural language processing.
- **Gather and Preprocess Data**: Gather the data that will be used to train the machine thinking algorithm. This data may need to be preprocessed to ensure that it is in the correct format and contains the necessary information.
- **Choose a Machine Thinking Algorithm**: Select the machine thinking algorithm that is best suited to the problem you are trying to solve. This could involve using a pre-built algorithm or developing your own.
- **Train the Algorithm**: Train the machine thinking algorithm using the data that you have gathered. This may involve using techniques such as supervised learning, unsupervised learning, or reinforcement learning.
- **Test and Refine the Algorithm**: Test the machine thinking algorithm using a validation dataset to ensure that it is accurate and effective. Refine the algorithm as necessary to improve its performance.
- **Deploy the Application**: Deploy the machine thinking application in a production environment, where it can be used to solve real-world problems.

Developing machine thinking applications requires a combination of domain-specific knowledge, programming skills, and expertise in machine thinking algorithms and techniques. In addition to this, it is equally important to understand different security issues in the context of machine thinking application development. Consider the scenario where the data used for training and testing machine thinking model is non-collocated, i.e., the data is available at different geographical locations. In critical use cases like healthcare, electronic health records are available at different cloud locations, federated learning approach [12] is the recommended solution. Similarly, there are several vulnerabilities which need to be considered for developing machine thinking applications. Security issues like authentication, access control, data integrity, trust are other important issues which require end-to-end solution by design instead of building a security layer on the top of the application. Key security issues are listed below:

- **Adversarial Attacks**: Adversarial attacks are a type of cyberattack that are specifically designed to target machine learning systems. These attacks involve manipulating the input data used to train the machine learning algorithm in order to cause it to make incorrect or unintended decisions.
- **Data Privacy**: Machine learning algorithms rely on large amounts of data to make decisions, and this data can contain sensitive or private information. If this

data is not properly secured, it could be accessed by unauthorized individuals or organizations.

- **Bias**: Machine learning algorithms can sometimes exhibit bias, which can result in unfair or discriminatory outcomes. This can be a security concern if the algorithm is used to make important decisions, such as in the criminal justice system.
- **Malware**: Malware can be used to compromise the security of machine learning systems, either by infecting the underlying hardware or by manipulating the input data used to train the algorithm.

The security concerns related to machine thinking are complex and multifaceted and require careful consideration and planning in order to effectively mitigate. Organizations that are developing machine thinking applications should take a proactive approach to security, implementing appropriate safeguards and testing their systems for vulnerabilities on a regular basis.

4.5 Summary

Machine thinking refers to the ability of machines to simulate human cognitive processes, including perception, reasoning, decision-making, and problem-solving. The concept of machine thinking has been the subject of much research and debate, with pioneers such as Alan Turing making significant contributions to the field. Today, machine thinking is a key component of artificial intelligence and machine learning and is being used to develop a wide range of applications in fields such as healthcare, finance, and transportation. However, there are also concerns related to the security and ethical implications of machine thinking, and these issues are being actively explored by researchers and policymakers. This chapter presents these issues along with the human and computer decision-making process. Application development in machine thinking and related key security issues are also presented and discussed in this chapter.

References

1. Churchland PM, Churchland PS (1990) Could a machine think? Sci Am 262:32–37
2. Wu X, Weng J (2021) On machine thinking. In: Proceedings on international joint conference on neural networks. Shengzhen, China, pp 1–8
3. Carbone MR (2022) When not to use machine learning: a perspective on potential and limitations. MRS Bull 47:968–974
4. Jarrett D, Stride E, Vallis K, Gooding MJ (2019) Applications and limitations of machine learning in radiation oncology. Br J Radiol 92(1100):20190001
5. Fazi MB (2019) Can a machine think (anything new)? automation beyond simulation. AI Soc 34:813–824
6. Johnson-Laird PN (1993) Human and machine thinking. Erlbaum, Hillsdale, N J

7. Tirado-Olivares S, Navío-Inglés M, O'Connor-Jiménez P, Cózar-Gutiérrez R (2023) From human to machine: investigating the effectiveness of the conversational AI ChatGPT in historical thinking. Educ Sci 13:803
8. Weng J (2020) A unified hierarchy for ai and natural intelligence through auto-programming for general purposes. J Cogn Sci 21(1):53–102
9. Busemeyer JR, Diederich A (2010) Cognitive modeling sage
10. Sun R (2001) Introduction to computational cognitive modeling. In: The Cambridge handbook of computational psychology (1st edn), Sun R (ed). Cambridge University Press, pp 3–20
11. Heathcote A, Brown SD, Wagenmakers EJ (2015) An introduction to model-based cognitive neuroscience 25–48. An introduction to good practices in cognitive modeling, an introduction to model-based cognitive neuroscience. Springer. https://doi.org/10.1007/978-1-4939-2236-9_2
12. Dixit SR, Mahalle PN, Shinde GR (2022) Rethinking data integrity in federated learning: are we ready? In: 2022 IEEE international women in engineering (WIE) conference on electrical and computer engineering (WIECON-ECE). Naya Raipur, India, pp 84–88

Chapter 5
Cognitive Computing

5.1 Internet of Behavior and Behavior Analysis

Internet of behavior or IoB is a multidisciplinary research area involving behavioral data analytics of humans and use of technologies to build innovative solutions. Internet of Things or IoT is an emerging field with the use of smart devices and the Internet to improve our lives. IoT mainly focuses on smart technologies and data analytics from the data gathered. IoB is an extension of IoT with interpretation of user interactions for informed decisions in the future. IoB attempts to convert gathered data and info into knowledge and wisdom which is as represented in Fig. 5.1.

Considering the increasing use of devices and the internet, huge data will be generated by millions of people around the world. The need for efficient big data analytics tools to handle this data and extract useful information from it is a challenge [1]. The potential use of 6G wireless communication services to revolutionize the systems involving customer services and applications are explored with possible applications in domains such as healthcare, autonomous driving, space technologies, and industries [2]. The potential use in wearable devices, health, supply chain, agricultural, and smart cities including homes, vehicles, devices provide enormous possibilities of using IoT for betterment. The potential challenges include storage of data, rules and regulations involving usage of data, privacy and security issues [3]. The ultimate objective of it is enhancing human life and better living.

Humans are evolving over the years and developed the skills and thought process through experience and knowledge. Artificial intelligence is posing a serious challenge to human capabilities with vast data processing and decision-making abilities. It attempts to mimic human behavior and attempts to build models based on the available knowledge and data. It was presumed that we use hardly 10% of the potential of our brain [4]. Neuroscientists, psychologists, and scientists have varied opinions about evolution of our brain and development [5]. Magnetic resonance imaging (MRI) has improved our understanding of the brain functioning in the last few years. Brain wide association studies (BWAS) involves modeling associations between

M. R. Velankar et al., *Cognitive Computing for Machine Thinking*, Innovations in Sustainable Technologies and Computing, https://doi.org/10.1007/978-981-97-0452-1_5

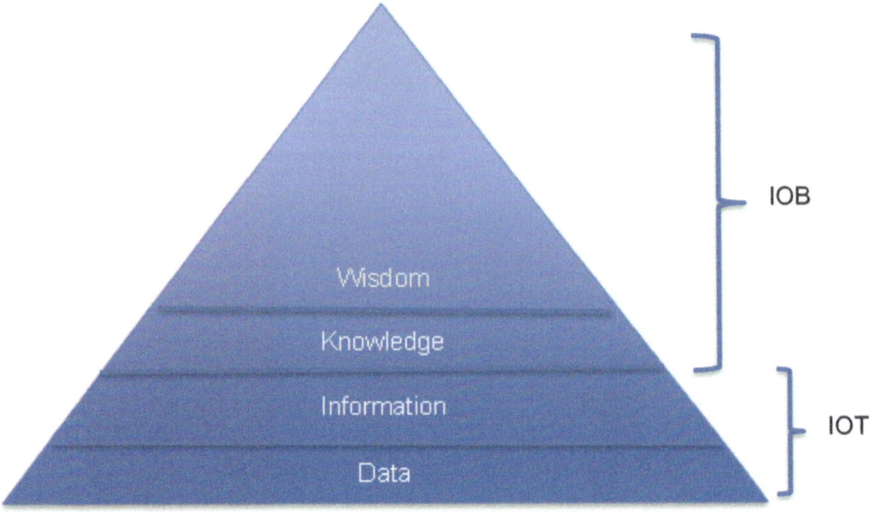

Fig. 5.1 IoT and IoB

inter-individual differences in brain structure and complex cognitive systems. Recent study shows BWAS modeling requires samples with thousands of individuals [6]. So, it is still a long way from complete understanding of the brain and associated behavioral analysis.

Some questions raise a deeper understanding about the human brain and behavioral analysis such as,

1. Is our brain basically a pattern recognition system?
2. What is the natural way to think?
3. What is the natural and apt expression of thought?
4. Can we really predict the behavior totally?

Behavioral analysis involves the study of psychological aspects of humans and non-humans. Non-humans' behavior is typically based on natural instincts and needs. For humans, in the majority of cases it is assumed that they make rational decisions based on the situation and previous experience, if any. However, in practice, it is observed that the decisions are influenced by the emotions and perception associated with the facts. Sentiment analysis based on human face or voice, or text provides certain clues which can be useful to model possible behavior for IoB. Similarly using different devices such as mobiles or Internet enabled devices can provide user's behavioral patterns, interests, and preferences to model them.

Behavioral psychology plays a crucial role in modeling the users which can be used as an effective tool for marketing, advertising recommending, and ultimately increasing the business for the organization. Thus, IoB can be useful for

- Analysis of buying pattern of the users.
- Understand the user interactions with devices, products, and services user.
- Developing psychological and behavioral models for accurate predictions.
- Improve user engagement and building strategy for sales.

Other side or possible disadvantages are privacy issues, breaching personal info, security issues, and possible cybercrimes. These issues need to be effectively handled in IoB during the implementation.

5.2 Data Points

Thea points are discrete units of info and in IoB they are the user facts which can be tracked while the user is accessing devices or the Internet. Data points of interest in IoB are the part of the user journey tracked to analyze the data about user behavior.

Customer journey map (CJM) is a useful tool used by organizations which attempts to find opportunities for improving customer relations and business ultimately. It was observed that most CJM has flaws as not all customers have the same touch points and importance of touch points varies for individuals [7]. Touch points are all possible interactions by customers for specific products or services. The touch points are typically divided into three sections before availing, during availing, and post availing of the service. These touch points can be used for developing data-driven marketing strategy [8]. CJM can be used for AI and big data framework for improving marketing and enhancing user experience [9]. An effective service innovation was proposed for e-commerce platforms till the last mile delivery of products using customer journey mapping [10].

CJM are used to bridge the gap between offerings by organization and expectations by the customers. They help to decide and choose value driven decisions and effective branding with consistent experience throughout the customer journey. Properly developed CJM has the following advantages.

- Easy to use and get insights of customer interactions.
- Helps to align different verticals or teams within the organization for the common vision and mission.
- Effective implementation of the branding strategy.
- Improve customer retention, engagement, experience, and fast conversion.
- Build strategies for better profitability.

Typical conventional steps followed for creating CJM are:

1. Decide a particular goal for creating CJM.
2. Create persona and identify customer goals.
3. Focus on specific crucial customer segment.
4. Identify all possible touch points.
5. Decide suitable type of CJM.
6. Search and select the resources and tools needed.

Fig. 5.2 Customer journey map

7. Experience the journey yourself to identify the crucial parts.
8. Modify the CJM to address the pain points of customers.

A sample customer journey map for online shopping is as shown in Fig. 5.2 from motivation to make payment to purchase goods.

The CJM is an effective visual tool to identify touch points. Data points are like touch points but with different objectives. The objective of data points is to capture the user behavior whereas the touch points are used to improve user experience. In online shopping examples, data points are captured in user tracking with browsing history by Internet service providers or a shopping portal. The sample data points for the use case of online shopping are as shown in Fig. 5.3. The representative data points are shown to understand the terminology data point with respect to user behavior analysis.

The data points are used to understand customer behavior and create a profile representing individuals or persona to create similar customers. Individual profiles are useful in personalized recommendations and persona is useful for collaborative filtering driven recommendations.

Systemic user experience approach is proposed based on CJM and personas for better designing the product or services [11]. Personas defines the customer goals and characteristics which are useful in designing products and services to match and at times exceed expectations. Data-driven persona development will be an efficient tool to capture the customer characteristics in a quantitative manner [12]. It helps to identify common interests with possible likes and dislikes.

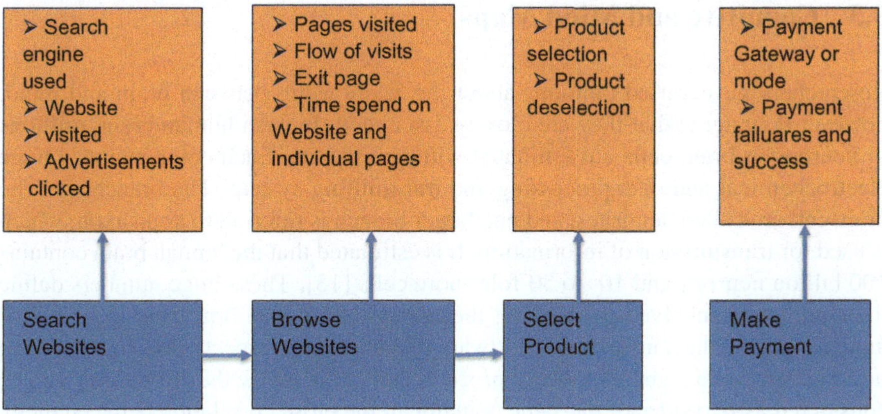

Fig. 5.3 Sample data points in customer journey map

Figure 5.4 represents a sample persona of the entrepreneurs with specific requirements from the mentoring app from Wadhwani foundation. It shows typical personal characteristics, goals, frustrations, and motivating factors of this customer segment. It helps to address important issues while building the app as a service for entrepreneurs. Persona shows the characteristics of the common behavioral patterns identified. These are outcomes of the thought process matured over a period by individuals. It can have deviations or changes over time depending on experience. The thought process can be represented using mind maps and related associations.

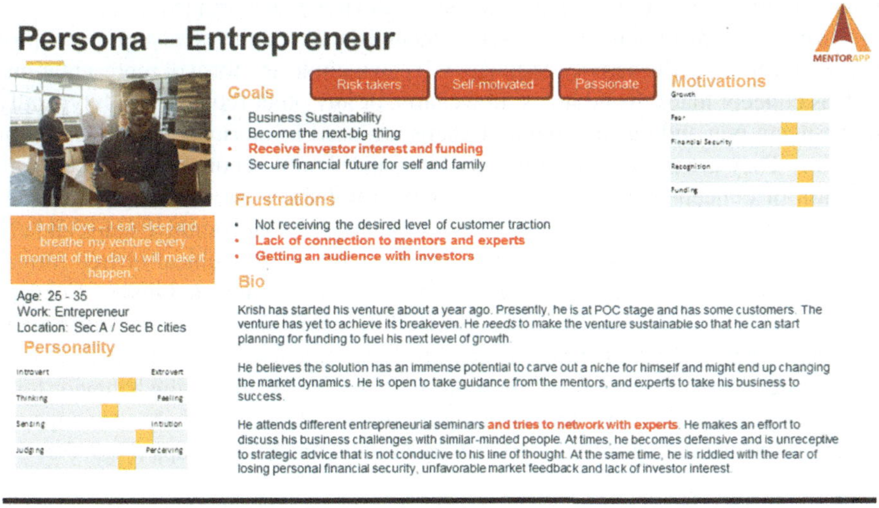

Fig. 5.4 Persona of entrepreneur

5.3 Cognitive and Mind Maps

Researchers have varied opinions about the relationship between brain and mind, however they agree that they are closely associated. In each human brain, millions of neurons or brain cells are estimated with complexity. Each cell contains a huge electrochemical and data processing and transmitting system. The branches of the brain cell are called dendrites, and one larger branch is referred to as an axon, which is used for transmission of information. It is estimated that the human brain contains 100 billion neurons and 10- to 50-fold more cells [13]. These huge numbers define the complexity involved in studying the human brain. The brain and the mind of humans are like the hardware and software of computers, respectively. The mind can be considered as a complex thought process happening inside the brain. They can be considered as two sides of the same coin due to the close correlation between them.

"Relation and interconnection between the brain and mind has been one of the most mysterious and still unresolved problems for the last two thousand years" [14]. Neuroimaging research related to mind revealed involvement of 13 regions in the brain for the spontaneous thoughts generated [15]. "A deeper investigation of more fleeting, spontaneous forms of cognition will be necessary to move cognitive neuroscience toward a neurobiological understanding of higher mental functions like creativity and imagination" according to authors. The cognitive process of thinking can be represented in visual forms for better understanding. Various visualization tools are used to represent the thought process in the human mind. Multilevel fuzzy cognitive maps [16] are proposed for opinion mining to evaluate public opinions. Knowledge representation using cognitive maps acts as a useful framework to represent complex info associated with brain-mind [17]. Cognitive maps are considered as an effective tool to organize knowledge for flexible behavior [18].

Cognitive maps in general represent a mental model of the person or a group of persons for the specific activity or concept. It can include any form of representations such as concept maps, listed items, flowcharts, or any other representation typically drawn using pen and paper. Cognitive maps are used for various tasks in different domains. They are not restricted by any form and flexible enough to represent any situation or concept. Cognitive maps provide a holistic view of the system under observation. Mind maps and concept maps are derived from cognitive maps and are structured representations. Mind map is represented in hierarchical or tree format whereas concept map is represented using graph or interconnected nodes. Concept maps are used to show relationships between different concepts whereas mind maps focus on a single concept. In practice, cognitive maps are used widely to represent the entire system and they are suitable to represent the mind. Considering behavior for a specific task, mind maps provide better representation. These tools help us to improve cognitive understanding using cognitive, mind and concepts = maps (Figs. 5.5, 5.6 and 5.7).

Mind map has one central concept or root of the tree with branches representing subtopics. It explores a single process or concept in detail with a detailed view.

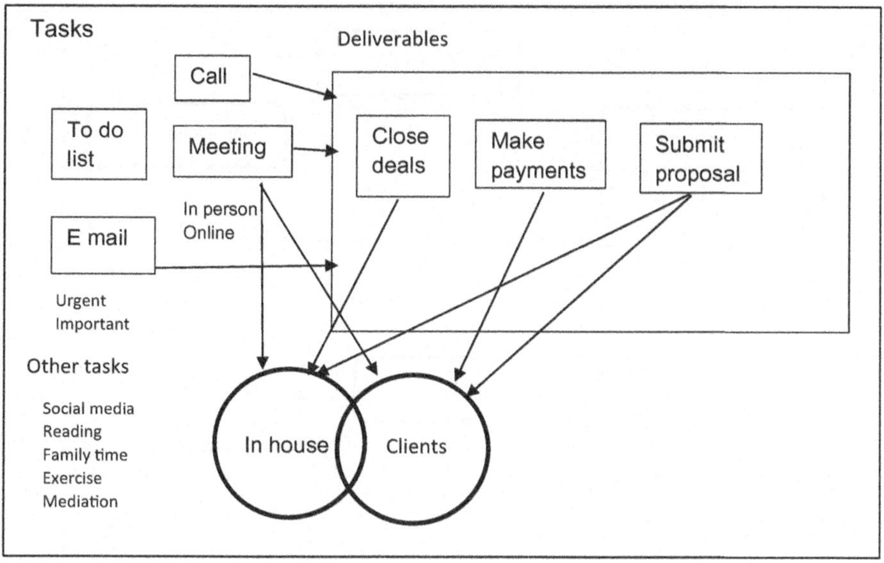

Fig. 5.5 Cognitive map

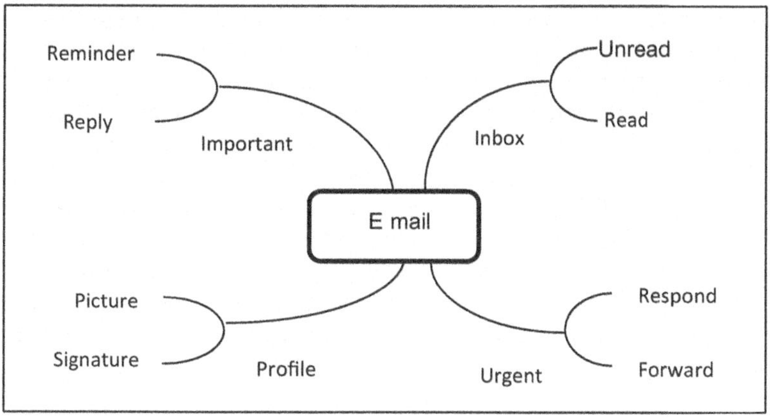

Fig. 5.6 Mind map

Mind mapping representation was effectively used for teaching programming skills to students at UG level [19].

Mind map attempts to model the thought process of the consumer with the use of important data points to model individuals.

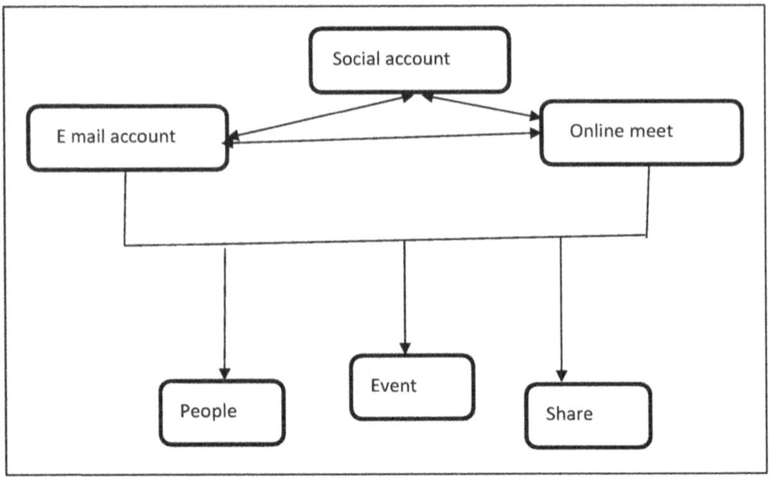

Fig. 5.7 Concept map

5.4 Programming for Mind

Mind programming is referred to as ways and techniques to control thoughts in psychology. It is used to make an affirmative positive tuning of the mind. Programming for the mind here refers to programming machines to think like humans and embed behavioral characteristics using the mind maps. Mind maps of humans will help to build predictive behavior model for the specific task or concept.

Various computing techniques are being used to train machines for the various tasks. Artificial intelligence is the broad area covering building overall intelligence in the machine. Machine learning algorithms, especially deep learning, are making remarkable progress and at times they are proving to be better than humans at performing certain tasks. Considering the need for huge data analytics in certain applications, machines have proved their utility. Deep learning architectures involving neural networks attempt to model neuron structures like the brain. The training of models with techniques such as supervised, unsupervised, semi-supervised, imitation and reinforcement are typically data driven. Learning with examples or labels helps machines to learn faster than without any examples or labels.

Programming for mind maps is modeling mind maps in machines. Mind map is represented as a tree with a central concept, which can be considered as a task to be done or a situation to be handled. The root typically refers to the task or situation and branches refer to the dependencies or allied info to be used for performing the task or resolving a situation or deciding. A sample mind map for the situation and allied info is shown in Fig. 5.8.

The possible inferences humans can derive from the available info are referred as knowledge gained which can be used to resolve the situation or take appropriate

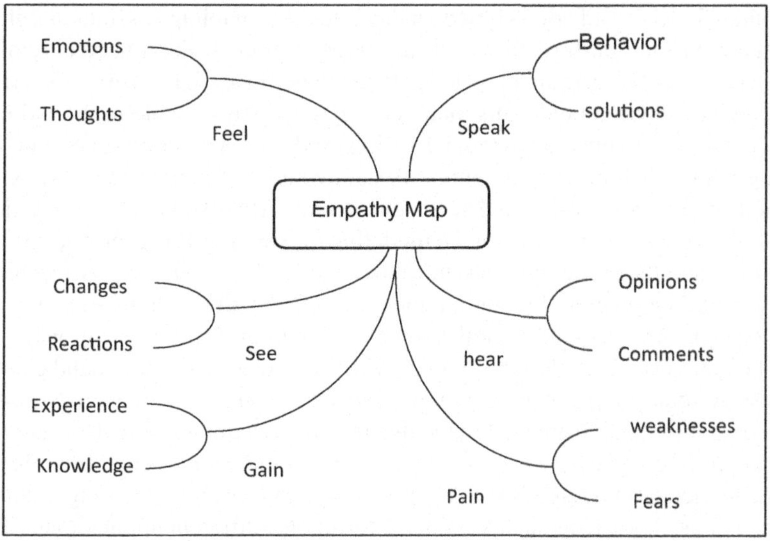

Fig. 5.8 Mind map for empathy

decisions. Typically, machine decisions are based on the facts available and are based on some objective function to optimize. Mind maps can add angles such as emotional state or social angle. Typical behavioral decisions are not only based on experience but also on the current mind state and involved entities. Personal affections, like/dislike, biases, perceptions affect human decisions considerably.

The objective of programming for the mind is to model the thought process of an individual or group of individuals. The cognitive abilities of humans in machines will make it more suitable for specific tasks.

5.5 Modeling Cognitive Thinking

Cognitive computing is a branch of AI that attempts to create systems capable of mimicking human cognitive functions, such as learning, reasoning, problem-solving, and understanding natural language [20]. Their objective is to effectively process and analyze extensive volumes of unstructured data to facilitate decision-making and offer insights, like the cognitive processes of human beings.

Cognitive computing systems depend on datasets consisting of structured and unstructured data types, including text, images, audio, and video data [21]. Such systems leverage machine learning methodologies to extract patterns, relationships, and knowledge from this data. Machine learning algorithms have the capacity to autonomously learn and adjust based on data, thereby enabling continuous enhancement of the system's performance over time. Furthermore, in the context of reinforcement learning, agents acquire knowledge through iterative trial and error experiences within a given environment. This methodology proves invaluable in the training of cognitive systems to make sequential decisions and optimize actions over time. Deep learning neural network architectures like convolutional neural networks (CNNs) designed for image analysis and recurrent neural networks (RNNs) for sequence data, have extensive application in the modeling of cognitive functions such as image recognition and the processing of natural language. NLP models allow machines to comprehend and generate human language. It enables the systems to engage with users through interfaces in natural language. Common NLP applications that are developed and used include sentiment analysis, language translation, and chatbots.

These systems are engineered with the capacity to engage in deductive reasoning and inference-making, from the knowledge they have acquired from data. This capability permits them to derive conclusions, address complex problems, and offer valuable recommendations. Symbolic AI approaches and machine learning, attempt to capture and use knowledge in a structured format for different applications [22].

To mimic human cognition, cognitive computing systems frequently combine knowledge representation techniques. This involves structuring information in a way that the system can reason about it effectively. Knowledge graphs and ontologies are commonly used for it [23]. Both knowledge graphs and neural networks serve as essential tools for representing and reasoning about complex relationships and knowledge structures. This makes them well-suited for cognitive computing tasks that require understanding of semantics. Additionally, the incorporation of multiple machine learning models into ensembles holds the potential to enhance the overall performance and flexibility of cognitive computing systems.

Cognitive computing systems hope to achieve context-awareness, signifying their ability to consider the situational context for making decisions or providing responses [24]. Proficiency in understanding context is key for nurturing interactions and solving problems in a manner like human-like intelligence. Cognitive computing models particularly distinguish themselves in discriminating patterns and finding anomalies within data. This attribute proves especially advantageous in applications such as fraud detection, image and speech recognition, and predictions. These systems are designed to collaborate with humans, thereby enhancing their capabilities as opposed to displacing them. They are intended toward aiding in complex tasks, furnishing recommendations, and reinforcing the decision-making process.

Cognitive computing continues to evolve, and researchers are exploring innovative approaches and techniques to make these systems more intelligent, adaptable, and human-like in their capabilities.

5.6 Cognitive Analytics

Cognitive analytics represents a highly advanced branch of analytical methodologies, fusing artificial intelligence (AI), machine learning, natural language processing (NLP), and various cognitive computing technologies. It revolves around the imitation of human cognitive processes, with the objective of deriving deeper and more meaningful insights from data [25].

The data obtained from the Internet of Behavior (IoB) serves as a large source of real-world behavioral data. Cognitive systems leverage this data to attain a more thoughtful understanding of how individuals interact with both the physical and digital domains, enabling a better understanding of human decision-making and responses. Cognitive analytics combines data from diverse sources, including both structured and unstructured data, to construct an all-inclusive perspective on information. This data integration plays a pivotal role in enabling cognitive systems to appreciate the context, considering aspects such as time, location, user conduct, and external events, all of which are essential for determining relevance.

Cognitive analytics extends its reach beyond mere correlation identification, digging into the identification of causal relationships within data. This capability empowers organizations to interpret the factors influencing specific outcomes. The system further showcases its ability in providing real-time insights through the continuous analysis of streaming data and automating decision-making by integrating predefined rules and learned patterns [26]. Such functionality proves particularly indispensable in scenarios requiring instant and data-driven decisions, such as in autonomous vehicles or algorithmic trading.

AI systems enhanced by cognitive analytics introduce a range of personalized features, recommendations, and tailored experiences. By precisely analyzing user behavior and preferences, these systems customize their responses to individual users, heightening user satisfaction and engagement. It's crucial to emphasize that cognitive analytics is not designed to replace human capabilities but, rather, to augment them. AI systems incorporating cognitive analytics extend support to human experts, offering insights, recommendations, and data-driven assistance, ultimately fostering more effective and efficient decision-making processes.

5.7 Underlying Technologies

Cognitive computing relies upon a diverse collection of foundational technologies aimed at matching human cognitive functions and providing intelligent capabilities. These technologies encompass a wide spectrum of domains, driven by the multifaceted nature of cognitive computing requirements.

At its core, machine learning techniques play a pivotal role, including supervised learning, unsupervised learning, imitation learning, and reinforcement learning. These algorithms provide the foundation for systems to acquire knowledge from data, identify patterns, and make informed predictions or decisions based on their learning. Within this domain, deep learning, a subset of machine learning involving artificial neural networks with multiple layers, is gaining significant attention. Particularly, transformative models like bidirectional encoder representations from transformers (BERT) and generative pretrained transformer (GPT) have provided groundbreaking advancements in natural language processing (NLP) tasks. Their applications extend to tasks such as language understanding, yielding promising outcomes. This is particularly crucial in applications related to text analysis, sentiment analysis, language translation, chatbots, and text summarization [27].

Computer vision technologies empower machines to interpret and understand visual information extracted from images or videos [28]. This proves crucial for tasks like object recognition, facial recognition, and image classification. Object detection algorithms play a key role in locating and identifying objects within images or video frames. Widely used techniques include Faster R-CNN, You Only Look Once (YOLO), and Single Shot MultiBox Detector (SSD). For semantic segmentation tasks, deep learning models like U-Net and Fully Convolutional Networks (FCNs) are deployed. Additionally, 3D computer vision tackles the estimation of the three-dimensional structure of objects or scenes from 2D images or video, with techniques encompassing stereo vision, structure from motion (SfM), and LiDAR-based methods. The importance of distributed computing and edge computing technologies emerges in real-time or near-real-time processing of visual data. It reduces latency and enhances efficiency in computer vision applications. In light of the growing prominence of deep fake technology, there is a pressing need for algorithms and technologies that can identify manipulated or artificial media content [29].

Semantic web technologies, represented by Resource Description Framework (RDF) and Web Ontology Language (OWL), facilitate the encoding of data with meaningful semantics, easing the process of understanding, and reasoning for machines [30]. Given the substantial volume of data handled by cognitive systems from diverse sources, technologies for data integration, data storage, and big data processing, such as Hadoop and Spark, are instrumental for data handling and analysis. User interface design technologies include voice and gesture recognition, touch interfaces, and augmented or virtual reality, fostering natural and intuitive interactions between humans and cognitive systems.

Explainable AI (XAI) technologies are pivotal in enhancing transparency and comprehensibility of AI and cognitive systems, offering explanations for their decisions and actions [31]. This aspect is vital for building trust and ensuring accountability. Considering the sensitive data handled by cognitive systems, cybersecurity technologies including encryption, authentication, and anomaly detection are critical in safeguarding against threats. Privacy-preserving techniques are also of significant importance in protecting user data. Quantum computing, an emerging technology with the potential to address complex problems and perform calculations at a scale exceeding that of classical computer systems, represents another dimension of advancement [32].

These foundational technologies continue to evolve and are normally combined in various ways to develop cognitive computing systems capable of accomplishing tasks that were once exclusively attributed to human intelligence.

5.8 Expected Capabilities in the System

The capabilities of cognitive systems play a pivotal role in their ability to meet complex real-world challenges. These capabilities make cognitive systems with human-like cognitive functions, enabling them to adapt to ever-changing environments and find application in various domains.

- Adaptive Learning: Cognitive systems must continually learn and adapt. They should be capable in analyzing new data, recognizing patterns, and refining their models and responses over time. This adaptability proves crucial in dynamic environments where new information and scenarios constantly emerge. Techniques like reinforcement learning, transfer learning, and unsupervised learning empower cognitive systems to autonomously enhance their performance and expand their knowledge base.
- Common Sense: Cognitive systems must possess the capacity for commonsense reasoning, allowing them to make logical inferences and deductions about the world. Integrating knowledge graphs, ontologies, and semantic understanding can foster more human-like reasoning abilities.
- Context Awareness: Cognitive systems need to be context-aware to offer relevant and meaningful interactions. They should consider a wide array of contextual factors, such as user history, location, time, and environmental conditions, when rendering decisions or recommendations. Moreover, the ability to understand and integrate information from multiple modalities, including text, images, audio, and sensor data, is essential for a holistic understanding of data and more comprehensive problem-solving.
- Multi-modal Capabilities: Enabling cognitive systems to seamlessly process and integrate information from diverse modalities nurtures more human-like interactions. This integrates understanding spoken language, interpreting visual data, and responding through natural language or other communication channels. Effective cross-modal integration augments the system's versatility and utility across applications like virtual assistants, autonomous vehicles, and healthcare.
- Decision-Making: In dynamic environments, cognitive systems must make real-time decisions. Whether it's autonomous vehicles navigating complex traffic or healthcare systems responding to patient data, the ability to process and act on information swiftly is vital. Dynamic adaptation to changing conditions and unexpected events is another critical capability.

- Human–Computer Interaction: Cognitive systems should seamlessly collaborate with humans, serving as intelligent partners, or assistants. This entails natural and intuitive interfaces for human interaction, cooperative decision-making, and the ability to augment human capabilities in various tasks.
- Ethical Considerations: Ethical considerations should be at the core of cognitive systems. They should be designed to detect and mitigate biases, ensuring fairness and equity in their operations. Moreover, cognitive systems should prioritize user privacy, incorporating robust security, and data protection mechanisms. Compliance with ethical guidelines and regulations is crucial for building trust and fostering responsible AI systems.
- Explainable Systems: Cognitive systems need to be transparent and capable of explaining their decisions and recommendations to users. This not only fosters trust but also helps users understand the rationale behind the system's actions. Research into XAI techniques, including interpretable machine learning models and visualization tools, is crucial for achieving transparency and accountability.

The field of cognitive computing is in a state of continuous evolution, and as it progresses, it will face many more forthcoming challenges in tackling complex real-world problems. Addressing these challenges is crucial to unlocking the full potential of cognitive systems across various domains.

5.9 Summary

Internet of behavior or IoB has emerged as a discipline which attempts to extract knowledge and gain wisdom from the information and data gathered from different devices and the Internet. It can be viewed as an improved or extended version of IoT with more insights into user patterns. User info in the form of customer journey map provides useful touch points to enhance user experience. Data points in the customer journey maps act as a major source of info about user behavioral patterns. The knowledge gathered from the data points helps to build customer persona with predictive analysis. For efficient predictions tools such as mind mapping can be used for behavior analysis. Machines can be programmed to model the human mind with thought processes represented as mind maps.

Cognitive computing along with data analytics is a powerful tool in building AI systems that can process vast amounts of data, understand context, make predictions, and provide valuable insights. It enhances the cognitive capabilities of AI, making it more adaptable, intelligent, and useful in a wide range of applications.

These essential capabilities of cognitive systems collectively enable them to tackle complex challenges in fields such as healthcare, finance, education, and more. The future of cognitive systems relies on ongoing research and innovation, interdisciplinary collaboration, and a commitment to ethical AI principles to ensure responsible and beneficial deployment across various domains.

References

1. Laghari AA, Wu K, Laghari RA, Ali M, Khan AA (2021) A review and state of art of internet of things (IoT). Arch Comput Methods Eng 1–19
2. Nguyen DC, Ding M, Pathirana PN, Seneviratne A, Li J, Niyato D, Poor HV et al (2021) 6G internet of things: a comprehensive survey. IEEE Internet Things J 9(1):359–383
3. Kumar S, Tiwari P, Zymbler M (2019) Internet of things is a revolutionary approach for future technology enhancement: a review. J Big Data 6(1):1–21
4. Radford B (1999) The ten-percent myth. Skeptical Inquirer 23:52–53
5. Cesario J, Johnson DJ, Eisthen HL (2020) Your brain is not an onion with a tiny reptile inside. Curr Dir Psychol Sci 29(3):255–260
6. Marek S, Tervo-Clemmens B, Calabro FJ, Montez DF, Kay BP, Hatoum AS, Dosenbach NU et al (2022) Reproducible brain-wide association studies require thousands of individuals. Nature 603(7902):654–660
7. Rosenbaum MS, Otalora ML, Ramírez GC (2017) How to create a realistic customer journey map. Bus Horiz 60(1):143–150
8. Micheaux A, Bosio B (2019) Customer journey mapping as a new way to teach data-driven marketing as a service. J Mark Educ 41(2):127–140
9. D'Arco M, Presti LL, Marino V, Resciniti R (2019) Embracing AI and Big Data in customer journey mapping: from literature review to a theoretical framework. Innov Mark 15(4):102
10. Vakulenko Y, Shams P, Hellström D, Hjort K (2019) Service innovation in e-commerce last mile delivery: mapping the e-customer journey. J Bus Res 101:461–468
11. Bradley C, Oliveira L, Birrell S, Cain R (2021) A new perspective on personas and customer journey maps: proposing systemic UX. Int J Hum Comput Stud 148:102583
12. McGinn J, Kotamraju N (2008) Data-driven persona development. In: Proceedings of the SIGCHI conference on human factors in computing systems, pp 1521–1524
13. Helmuth L (2001) Glia tell neurons to build synapses
14. Reshetnikov MM (2017) Problem of relation between brain and mind in physiology, medicine and psychology. J Psychiatry Psychiatric Disorders 1(6):313–316
15. Fox KC, Spreng RN, Ellamil M, Andrews-Hanna JR, Christoff K (2015) The wandering brain: meta-analysis of functional neuroimaging studies of mind-wandering and related spontaneous thought processes. Neuroimage 111:611–621
16. Sánchez H, Aguilar J, Terán O, de Mesa JG (2019) Modeling the process of shaping the public opinion through multilevel fuzzy cognitive maps. Appl Soft Comput 85:105756
17. Peer M, Brunec IK, Newcombe NS, Epstein RA (2021) Structuring knowledge with cognitive maps and cognitive graphs. Trends Cogn Sci 25(1):37–54
18. Behrens TE, Muller TH, Whittington JC, Mark S, Baram AB, Stachenfeld KL, Kurth-Nelson Z (2018) What is a cognitive map? Organizing knowledge for flexible behavior. Neuron 100(2):490–509
19. Liu Y, Tong Y, Yang Y (2018) The application of mind mapping into college computer programming teaching. Proc Comput Sci 129:66–70
20. Modha DS, Ananthanarayanan R, Esser SK, Ndirango A, Sherbondy AJ, Singh R (2011) Cognitive computing. Commun ACM 54(8):62–71
21. Gupta S, Kar AK, Baabdullah A, Al-Khowaiter WA (2018) Big data with cognitive computing: a review for the future. Int J Inf Manage 42:78–89
22. Farrell R, Lenchner J, Kephart J, Webb A, Muller M, Erickson T, Gil D et al (2016) Symbiotic cognitive computing. AI Mag 37(3):81–93
23. Zheng P, Xia L, Li C, Li X, Liu B (2021) Towards self-X cognitive manufacturing network: an industrial knowledge graph-based multi-agent reinforcement learning approach. J Manuf Syst 61:16–26
24. Huet A, Pinquié R, Véron P, Mallet A, Segonds F (2021) CACDA: a knowledge graph for a context-aware cognitive design assistant. Comput Ind 125:103377
25. Handfield R, Jeong S, Choi T (2019) Emerging procurement technology: data analytics and cognitive analytics. Int J Phys Distrib Logist Manag 49(10):972–1002

26. Bratu S, Sabău RI (2022) Digital commerce in the immersive metaverse environment: cognitive analytics management, real-time purchasing data, and seamless connected shopping experiences. Linguistic Philos Investigations 21:170–186
27. Dai Z, Yang Z, Yang Y, Carbonell J, Le QV, Salakhutdinov R (2019) Transformer-xl: attentive language models beyond a fixed-length context. arXiv preprint arXiv:1901.02860
28. Voulodimos A, Doulamis N, Doulamis A, Protopapadakis E (2018) Deep learning for computer vision: a brief review. Comput Intell Neurosci
29. Westerlund M (2019) The emergence of deepfake technology: a review. Technol Innovation Manage Rev 9(11)
30. Allemang D, Hendler J (2011) Semantic web for the working ontologist: effective modeling in RDFS and OWL. Elsevier
31. Gunning D, Aha D (2019) DARPA's explainable artificial intelligence (XAI) program. AI Mag 40(2):44–58
32. National Academies of Sciences, Engineering, and Medicine (2019) Quantum computing: progress and prospects

Chapter 6
Business Cases

6.1 Overview

AI is a disruptive technology with far-reaching applications in many multidisciplinary areas. AI also raises philosophical and ethical issues that cross disciplinary boundaries, prompting interdisciplinary debates about its societal implications. AI's versatility and flexibility make it an essential tool in tackling complex problems and opportunities across diverse domains, driving innovation, and changing the way we confront multidisciplinary challenges.

AI and ML models are used in the fields of environmental science and sustainability. These models can be used to monitor the effects of climate change and natural disasters on the environment, as well as to conserve biodiversity. In addition, smart grids and intelligent energy management systems can be used to optimize energy consumption and distribution, helping to reduce carbon emissions and promote sustainability. On the finance and economics front, AI-driven algorithms can be used to analyze financial markets and identify trading opportunities. They can also be used to manage investment portfolios and make better financial decisions. On the economic forecasting front, AI can be used to process large amounts of economic data, which can be used to inform policy and strategy for governments and businesses. In education and learning, AI-powered educational tools provide personalized learning experiences, customizing content, and pacing to meet the needs of each student. In NLP, automated grading and feedback are provided to students, teachers, and researchers. In transportation and urban planning, AI improves autonomous vehicles' ability to drive safer and more efficiently, leading to fewer accidents, and congested roads. In traffic management systems, AI optimizes traffic flow and reduces commuting times in urban areas.

AI-powered art, music, and literature open up new possibilities for creative expression and cultural discovery. AI-powered design tools help architects, fashion

M. R. Velankar et al., *Cognitive Computing for Machine Thinking*, Innovations in Sustainable Technologies and Computing, https://doi.org/10.1007/978-981-97-0452-1_6

designers, and artists in their design processes. National security and defense: AI-powered systems help with threat detection, cybersecurity and intelligence analysis, improving national security.

AI-powered autonomous drones provide surveillance, AI and NLP are used in social sciences and psychology to analyze social media information to understand public opinion, trends, and crises. AI-powered chatbots and AI-powered virtual assistants provide mental health services and counseling. Space exploration AI is utilized in space missions to enable autonomous navigation and robotics, as well as data analysis, to explore the cosmos in greater depth and complexity.

AI evokes deep philosophical questions and moral dilemmas, provoking interdisciplinary debates about AI's social impact, discrimination, justice, and interactions between humans and machines. In each of these interdisciplinary settings, AI accelerates innovation, improves decision-making and deepens our understanding of the big picture. Its capacity to analyze large datasets, identify trends and make predictions goes beyond the boundaries of traditional disciplines, making it an invaluable tool for tackling multifaceted issues and opportunities in our ever-connected world [1].

6.2 Internet of Things (IoT)

AI has transformed the IoT ecosystem by allowing connected devices to be more intelligent and responsive [2–5]. For example, AI-powered IoT devices in smart homes, such as smart thermostats or smart lighting systems, can learn user behavior and adjust to create more comfortable and energy efficient environments. For industries, AI-powered predictive maintenance systems can predict equipment failures and reduce costly downtime. AI also has a positive impact on environmental monitoring. IoT sensors and intelligent AI algorithms can detect pollution levels and ensure natural resources are conserved. AI also improves supply chain management, allowing IoT data to be used to optimize inventory, monitor traffic conditions, and optimize logistics.

- Data analytics and insights: IoT is a network of sensors and devices that generate billions of data points every day. These data points are analyzed in real-time using AI algorithms, especially ML and DL. AI can detect patterns, anomalies, and trends in the data and provide actionable insights to help you make better decisions. For instance, in the industrial IoT domain, AI-powered predictive maintenance can predict equipment failure and optimize maintenance schedules to reduce downtime and costs.
- AI-driven predictive maintenance: Models leverage real-time and historical data from IoT sensors to identify when equipment is most likely to malfunction. This proactive approach reduces downtime and maintenance costs for industries like manufacturing, aviation, and energy.

- Anomaly detection: It is the detection of abnormal behavior or anomalies within IoT data by AI algorithms. These anomalies can be a sign of a security breach or system failure. In the IoT security context, AI-based anomaly detection helps protect networks and devices against cyber threats.
- Energy efficiency: IoT devices are installed in smart networks and smart buildings for energy monitoring and control. AI helps optimize energy consumption by analyzing data from IoT devices, real-time adjusting heating and cooling, lighting, and other systems to reduce energy consumption and costs.
- Voice assistants: IoT devices use natural language processing (NLP) and AI to understand and answer voice commands, which improves user interactions and allows for voice controlled smart home appliances and devices.
- Computer vision: In IoT, AI-powered computer vision is used to identify, track, and monitor objects. AI-powered IoT cameras and sensors can identify intruders, track traffic, and improve security in public places and homes.
- AI-driven autonomous vehicles: unmanned aerial vehicles (UAVs), and robotics often use IoT sensors to collect data and navigate. These systems rely on AI algorithms to process sensor information, make decisions in real-time, and navigate. IoT devices embedded in transportation vehicles gather information about traffic conditions, fuel usage, and vehicle operation. AI algorithms then use this information to optimize routes, lower fuel costs, and predict arrival times more precisely, resulting in more efficient transportation and reduced operating costs.
- Supply chain optimization: AI and IoT combine to monitor inventory levels, monitor transportation conditions (such as temperature-sensitive goods) and provide real-time insights on the movement of goods to improve supply chain operations.
- Real-time visibility: IoT sensors are installed in manufacturing facilities, distribution centers, and transportation vehicles throughout the supply chain. IoT sensors gather real-time information about where, what, and when goods and assets are in real-time. This information is processed by AI-driven analytics, allowing supply chain managers to view the entire process in real-time, helping to improve decision-making, reduce delays, and minimize inventory holding costs, e.g., predictive analytics can use weather data or GPS data from IoT devices to predict delays due to bad weather. With predictive analytics, you can take proactive steps, such as redirecting shipments or rescheduling production schedules to reduce the risk of supply chain issues. AI-driven demand forecasting models use IoT data to accurately forecast product demand. This optimizes inventory levels, reducing the likelihood of stock spikes or shortages. IoT sensors can also monitor storage conditions, including temperature and humidity levels, and notify stakeholders when conditions are out of range.
- Environmental monitoring: IoT sensors combined with AI monitors and analyzes environmental data, such as air and water quality and weather patterns, as well as wildlife behavior. This data aids in environmental protection and disaster management.

6.3 Healthcare

AI is revolutionizing healthcare with its capacity to analyze large datasets and make predictive predictions that enhance patient care. AI-driven algorithms look at medical images, such as X-rays or MRIs, to help radiologists diagnose diseases and abnormalities in patients [6, 7]. Drug discovery is another area where AI is revolutionizing medicine. AI helps identify drug candidates faster and assess their effectiveness, accelerating the development of novel treatments. Personalized medicine is also an area where AI excels. It uses patient data to develop personalized treatment plans and predicts disease risk factors to help improve patient outcomes. Healthcare IoT devices, combined with AI, allow remote health monitoring and early intervention, as well as streamlining administrative tasks such as scheduling appointments and medical billing.

AI is revolutionizing medical imaging, providing healthcare professionals with unprecedented capabilities. By analyzing complex medical images quickly and accurately, AI is changing the way we diagnose and treat various medical conditions. AI algorithms can identify anomalies, tumors, fractures, and more in X-ray, CT scan, MRIs, ultrasound images. They are a key benefit in early disease detection, enabling more effective interventions and improving patient outcomes. AI also automates tasks such as image segmentation and 3D reconstruction, as well as quantitative analysis, which streamlines workflows and improves the accuracy of treatment planning. In addition to image quality and differential diagnosis, AI's benefits extend to telehealth, where expert opinions can be more easily accessed in remote locations. In clinical trials and in research, AI makes it easier to analyze large-scale images, ensuring consistent and accurate data interpretation. As AI advances, its impact on medical imaging is set to revolutionize the field, empowering healthcare professionals to improve diagnostic precision, optimize treatment approaches, and ultimately, improve patient care.

AI is transforming disease diagnosis by leveraging its powerful data analysis, pattern detection, and predictive modeling capabilities. AI systems can process large amounts of patient information, including medical records, test data, and imaging images, to help healthcare professionals diagnose and treat a wide variety of diseases. One of the biggest benefits of AI in disease diagnostics is its ability to identify tiny patterns and anomalies that may otherwise go undetected by human clinicians. Whether it's early cancer detection in medical images or early detection of rare diseases in genomic data, AI's accuracy and efficiency are changing the diagnostic landscape. This not only results in faster and more accurate diagnoses, but also allows for early intervention and treatment, resulting in improved patient outcomes and life savings. As AI continues to develop and expand its use cases in healthcare, it has the potential to be a game-changer in the ongoing fight against disease, providing healthcare providers with improved diagnostic tools and patients with increased hope of early detection and successful treatment.

AI has revolutionized drug discovery by accelerating the discovery of drug candidates and improving our understanding of intricate biological systems. Machine learning and deep learning algorithms can analyze large datasets of biological and chemical information to predict how particular molecules interact with biological target molecules. This allows drug compounds to be identified in a fraction of the time and cost typically associated with drug discovery. AI also improves drug development pipelines, predicts drug safety profiles, and even rewrites existing drugs for new uses. Accelerating drug discovery at the early stages opens up new opportunities for treating rare and neglected diseases, as well as providing more personalized and tailored therapies. AI also encourages multidisciplinary teams of chemists and biologists, as well as data scientists, to work together to drive innovation and bring new treatments to market more quickly. As AI in drug discovery advances, it has the potential to solve some of the world's most pressing healthcare issues and improve the lives of millions of people around the world.

AI is leading the way in personalized medicine. AI algorithms can analyze an individual's unique genetic makeup and medical history, as well as lifestyle data, to create personalized treatment plans and interventions that are tailored to meet the specific needs of each patient. This is a major departure from traditional medicine, as it allows healthcare providers to create more effective treatment plans with fewer side effects.

AI-powered predictive models can predict how patients will respond to different therapies, helping doctors select the best treatment for a variety of conditions. For example, in oncology, AI helps doctors select the best cancer treatment for a patient based on a patient's genomic profile, increasing the likelihood of successful treatment. AI improves early disease detection by recognizing risk factors and predicting susceptibility to disease, allowing for prevention and early intervention. Not only does this approach improve patient outcomes, but it also reduces healthcare spending by preventing unnecessary treatments and side effects.

AI-powered personalized medicine is a game-changer in providing more accurate and efficient healthcare, tailored to each patient's unique genetic and clinical profile. As AI advances, its impact on personalized medicine will only grow.

Generative AI has a lot of potential in the healthcare space, but it also brings ethical and regulatory issues, especially when it comes to the authenticity and trustworthiness of generated data. Generated content must meet ethical standards and be used responsibly in the healthcare space.

6.4 Musicology

AI has transformed the music industry, improving the way we find, create, and listen to music. AI-powered recommendation systems, such as those found on popular streaming platforms such as Spotify, offer personalized music recommendations based on users' preferences and listening habits. AI can even write music, providing composers and artists with valuable tools. AI's audio analysis capabilities allow it to detect patterns, emotions, and copyright infringements in audio tracks. AI-powered virtual voice assistants, such as Siri and Alexa, understand voice commands associated with music playback and control to enrich the music listening experience [8].

AI is revolutionizing the way we compose music. It's the beginning of a new age where human artists collaborate with machines to create original music. AI-powered composition tools can create original music in almost any genre and style. AI algorithms analyze large libraries of music, finding patterns, harmony, and melody, which they use as a basis to create completely new compositions. Not only does AI help musicians and compositionists to come up with new ideas, but it also serves as an invaluable tool for content creators looking for royalty free music for different projects. AI is capable of producing music faster and more efficiently than humans. It can assist composers in overcoming creative obstacles, create background music for video games, commercials, and music videos, and even work with human performers to produce hybrid compositions. AI-generated music may raise questions about the place of human creativity within the arts, but it also provides new opportunities for creativity and experimentation. As AI-powered musical composition tools develop, they are set to revolutionize the music industry and provide powerful tools for professional musicians as well as amateur creators to express and explore music.

AI tools and algorithms are making it easier, more accessible, and more creative to produce music [9]. AI-powered systems can help music producers automate mundane tasks, like mixing and mastering, so they can spend more time working on their creative side. They can analyze audio tracks, change levels, use effects, and even recommend improvements in real-time, saving time, and energy. They can improve the overall quality of your sound recordings and help you achieve a more polished and professional sound.

In addition, AI can help create music content by helping to create melodies, harmonies, and even instrumentals. This helps musicians and producers during the ideation stage, giving them fresh ideas for their songs. AI also helps to make music production more accessible to more people. Even people with limited musical knowledge can use AI-driven software to create high-quality music.

AI is revolutionizing the way we create music content. Artists, content creators and musicians can use AI-powered content creation tools to create original music, lyrics, and even entire tracks across different genres and styles with ease and speed. AI algorithms analyze huge data sets of existing music to find patterns, melodies, harmonies and lyrics, which are then used as a foundation for creating new compositions.

In addition to helping musicians overcome obstacles and come up with new concepts, AI-powered content creation tools can be useful for content creators looking for royalty free, high-quality music for a variety of purposes, such as video content, podcast content, and advertising. They offer a convenient and cost-effective way to listen to music that matches the tone and style you want your content to have. AI can also adapt to your creative preferences, so you can customize the generated content to your liking. This allows for human artists to collaborate with AI to create unique and creative musical expressions.

6.5 Agriculture

AI-powered technologies are improving yields, reducing resource waste, and increasing sustainability in agriculture [10, 11]. Computer vision enabled by AI can detect diseases and pest infestations in crops, enabling targeted interventions and reducing crop losses. AI-powered livestock monitoring takes advantage of IoT devices and AI-powered analytics to ensure the welfare of animals and increase productivity. AI also helps to optimize the agricultural supply chain—from production to distribution—reducing waste and inefficiency.

Precision agriculture is the use of AI-powered technologies, such as drones and IoT sensors, to gather data on soil health, weather, and plant health. The data is then analyzed by machine learning algorithms to determine the best ways to plant, irrigate, and fertilize crops to maximize resource efficiency and yield. AI collects and analyzes large amounts of data, such as soil composition, weather, crop health, historical performance. This data allows farmers to make decisions based on data, like when to plant, when to irrigate, and when to harvest. Minimizing Resource Waste: Precision agriculture reduces resource waste by applying water, fertilizer, and pesticides precisely where and when they need to be applied. This lowers costs for farmers and reduces the environmental impact of agriculture.

Crop monitoring is the use of drones with cameras and AI algorithms that capture high-quality images of fields in real-time. AI can detect areas where pests, diseases, or nutrient deficiencies are present, allowing for early intervention to improve crop health and reduce risk. Machine learning predictive analytics machine learning models can be used to predict crop yields, market conditions, and other factors based on past and current data. This allows farmers to optimize their planting and harvest schedules to maximize profits and minimize risk.

Weed Identification Cameras and sensors powered by AI are installed on agricultural equipment, including tractors and unmanned aerial vehicles (drones), to capture real-time images of fields. AI-powered machine learning algorithms then analyze the images to identify crops and weeds. This allows for more accurate weed identification, while also reducing the risk of damaging desirable crops. AI controlled sprayers utilize the weed identification data to spray herbicides only on identified weeds. By reducing the amount of chemicals used, farmers reduce operational costs and damage to the environment. Robotic weed removal is the use of autonomous robots equipped with artificial intelligence technology to navigate fields and remove weeds without the need for human intervention. By using computer vision, these robots are able to identify and remove weeds with high levels of accuracy. This method of weed removal is an environmentally friendly alternative to the use of chemical herbicides, which can be harmful to the environment. AI provides valuable data on weed populations and weed distribution within fields, allowing farmers to make more informed decisions about weed management strategies and allocate resources more efficiently. AI-powered weed control systems work seamlessly with GPS and farm management software to map weed infected areas and plan targeted actions.

6.6 Education

AI is changing the way we learn, improving the way we teach, and improving the way we learn outcomes. Some of the main uses of AI in education include [12, 13].

Personalized learning: AI-powered educational platforms customize content and pace to meet the unique learning needs and learning preferences of each learner. This personalized approach increases engagement and knowledge retention.

AI-powered tutoring: Systems can provide students with instant feedback and support in areas such as math and language. They can spot areas where students are struggling and provide personalized assistance.

AI-powered automated grading and assessment: AI can automatically grade assignments, tests, and quizzes quickly and consistently. This saves teachers time and provides students with real-time feedback about their performance.

Virtual classrooms: These virtual classrooms are powered by AI and offer remote learning features such as video conferencing and screen sharing, as well as interactive whiteboards. With the help of AI, it is possible to track student engagement and participation and identify any potential problems.

NLP: NLP is used to create AI chatbots or virtual teaching assistants that answer students' questions and guide them through the coursework.

AI-powered content creation: It helps educators create engaging and informative content for their students, such as educational videos and quizzes, as well as study materials.

AI-powered early intervention helps educators identify students who are at risk of lagging behind or struggling in certain subjects, so they can offer targeted interventions to help them catch up. Language learning apps use AI to create personalized lessons, improve speech recognition, and improve language acquisition.

Accessibility: AI helps to improve the accessibility of educational materials for students with disabilities by enabling features such as text to speech, speech to text, and captions. Educational Research: AI helps to analyze educational data to detect trends, evaluate teaching effectiveness, and inform education policy and curriculum development. AI helps with resource allocation, from class scheduling and budgeting to predicting future enrollment.

Security: AI helps with security, from tracking cheating on online exams to keeping students safe in physical environments.

AI in education promises to improve the quality, accessibility, and effectiveness of learning. But it also raises ethical and privacy issues, particularly when it comes to the treatment of student data and fairness and transparency in an AI-driven education system. As AI technology advances, its influence on education is expected to increase, presenting both opportunities and challenges to educators, learners, and policy makers.

6.7 Summary

AI is changing the way we live, work, and play. In IoT, AI-powered analytics and predictive models extract insights from billions of data streams, streamlining operations, and improving security. In healthcare, AI helps with diagnosis, drug discovery, and personalized treatment plans to improve patient care. In agriculture, AI helps with precision, streamlining resource use, and improving crop management. In music, AI helps with composition, production and recommendation, encouraging creativity. In education, AI-powered personalized learning and intelligent tutoring, as well as virtual classrooms, transform teaching methods. All of these applications demonstrate how AI can improve efficiency, sustainability, and innovation across a wide range of domains.

References

1. Dwivedi YK, Hughes L, Ismagilova E, Aarts G, Coombs C, Crick T, Williams MD et al (2021) Artificial Intelligence (AI): multidisciplinary perspectives on emerging challenges, opportunities, and agenda for research, practice and policy. Int J Inf Manage 57:101994
2. Al-Turjman F (ed) (2019) Artificial intelligence in IoT. Springer
3. Merenda M, Porcaro C, Iero D (2020) Edge machine learning for ai-enabled iot devices: a review. Sensors 20(9):2533
4. Singh R, Srivastava S, Mishra R (2020) AI and IoT based monitoring system for increasing the yield in crop production. In: 2020 International conference on electrical and electronics engineering (ICE3). IEEE, pp 301–305
5. Ahuja K, Bala I (2021) Role of artificial intelligence and IoT in next Generation education system. Intelligence of things: AI-IoT based critical-applications and innovations, 189–208
6. Yu KH, Beam AL, Kohane IS (2018) Artificial intelligence in healthcare. Nature Biomed Eng 2(10):719–731
7. Shaheen MY (2021) Applications of artificial intelligence (AI) in healthcare: a review. ScienceOpen Preprints
8. Civit, M., Civit-Masot, J., Cuadrado, F., & Escalona, M. J. (2022). A systematic review of artificial intelligence-based music generation: Scope, applications, and future trends. *Expert Systems with Applications*, 118190.
9. Patil SS, Shelke S, Velankar M, Koul T, Rani V (2023) A systematic survey of approaches used in computer music generation. J Innovations Data Sci Big Data Manage 2(1):16–23
10. Liu SY (2020) Artificial intelligence (AI) in agriculture. IT Professional 22(3):14–15
11. Singh P, Kaur A. (2022) A systematic review of artificial intelligence in agriculture. Deep Learn Sustain Agricult 57–80
12. Holmes W, Tuomi I (2022) State of the art and practice in AI in education. Eur J Educ 57(4):542–570
13. Limna P, Jakwatanatham S, Siripipattanakul S, Kaewpuang P, Sriboonruang P (2022) A review of artificial intelligence (AI) in education during the digital era. Adv Knowl Executives 1(1):1–9

Chapter 7
Conclusion

7.1 Summary

Machine thinking enables machines and AI systems to process information, reason, and make decisions like human thinking. It involves the use of algorithms and computational models to simulate cognitive processes such as perception, learning, problem-solving, and decision-making. The need for machine thinking arises from the increasing complexity of tasks and the desire to automate them. Machines with the ability to think can analyze large amounts of data, identify patterns, and generate insights that can be used to make informed decisions. This has numerous applications across various fields, including healthcare, finance, manufacturing, and transportation.

Chapter one of this book presents an overview of natural and artificial intelligence. They are interconnected in various ways. Advancements in AI, particularly machine learning and neural networks, have contributed to progress in neuroscience. Machine learning techniques are used in natural language processing (NLP), enabling applications such as spam detection, machine translation, virtual agents, chatbots, and social media sentiment analysis. Cloud-based NLP platforms utilize machine learning to extract, analyze, and store text data. NLP technology, combined with AI, allows users with no prior experience in AI or NLP to utilize document analysis capabilities. AI, including machine learning and NLP, is an evolving landscape with each technology progressing along its own path. Generative AI algorithms, such as ChatGPT and DALL-E, have applications in various fields like medical imaging analysis and weather forecasts.

AI and cognitive computing are closely related fields that utilize artificial intelligence techniques to simulate human thought processes and enhance computing systems' capabilities. AI refers to the development of computer systems that can perform tasks that typically require human intelligence, such as speech recognition, decision-making, and problem-solving. It encompasses various techniques, including machine learning, natural language processing, and computer vision.

M. R. Velankar et al., *Cognitive Computing for Machine Thinking*, Innovations in Sustainable Technologies and Computing, https://doi.org/10.1007/978-981-97-0452-1_7

Cognitive computing, on the other hand, focuses on creating systems that can mimic human cognitive abilities, such as perception, learning, memory, and reasoning. It aims to develop computing models that can understand, analyze, and interpret complex data in a manner like human cognition. Cognitive computing utilizes AI techniques, such as machine learning and pattern recognition, to enable systems to learn from data, recognize patterns, and make informed decisions. It emphasizes human–computer interaction and aims to enhance collaboration between humans and machines. The goal of cognitive computing is to create systems that can understand and interpret unstructured data, such as text, images, and videos, and provide insights and recommendations based on this data. It combines AI algorithms with cognitive science principles to develop more accurate models of human thinking and response.

In summary, AI and cognitive computing are interconnected fields that leverage artificial intelligence techniques to enhance computing systems' capabilities and simulate human thought processes. Both fields contribute to advancements in various domains, including natural language processing, image recognition, and decision support systems and these details are presented and discussed in the chapter two of this book. This chapter also deals with phenomenon in human cognition, augmented intelligence along with the concepts of machine and human behavior.

The third chapter of this book deals with rethinking machine learning and deep learning. Rethinking machine learning and deep learning involves critically examining the existing approaches and paradigms in these fields and exploring new ideas and perspectives to enhance their effectiveness and address their limitations. This includes reconsidering the role of priors in Bayesian learning, exploring the application of machine learning in cardiac electrophysiology analysis and breast cancer diagnosis, and reevaluating the generalization capabilities of deep learning models.

Some researchers argue that understanding deep learning requires a rethinking of generalization, as traditional models may not fully explain the behavior of deep neural networks. Additionally, there is a need to reconsider the annotation granularity in deep learning algorithms to overcome shortcuts and improve performance. The deep learning revolution has led to the popularity of this approach in various applications, and it is important to understand its principles, potential, and integration with existing machine learning frameworks. Rethinking machine learning and deep learning involves questioning existing assumptions, exploring new methodologies, and seeking improvements to enhance the capabilities, interpretability, and generalization of these models.

Alan Turing's seminal paper "Computing Machinery and Intelligence" explores the question of whether machines can think. Turing argues that it is difficult to define thinking and machines in a precise manner, suggesting that the more appropriate question is whether machines can exhibit intelligent behavior. Machine thinking often relies on machine learning techniques, where algorithms learn from data to make predictions or decisions. Deep learning, a subset of machine learning, utilizes artificial neural networks to simulate the learning processes of the human brain.

Advancements in machine thinking have led to significant breakthroughs in various domains, including natural language processing, computer vision, and autonomous systems. Machine thinking has enabled machines to perform complex tasks, such as language translation, image recognition, and self-driving cars. However, there are ongoing debates and challenges in machine thinking, including the limitations of current AI systems in terms of explainability, transparency, and ethical considerations. Researchers are constantly rethinking and refining machine learning and deep learning approaches to improve their interpretability, generalization capabilities, and fairness.

Machine thinking refers to the development of machines capable of exhibiting intelligent behavior and cognitive processes. It involves the application of machine learning and deep learning techniques to enable machines to perceive, reason, learn, and make decisions in a manner like human thinking. Ongoing research aims to overcome challenges and enhance machine thinking capabilities and all these details are discussed in the fourth chapter of this book.

The Internet of behaviors (IoB) is a concept that involves the collection and analysis of data to understand and predict human behaviors. It combines data analysis, behavioral analysis, and technology to gain insights into individual and collective behaviors. IoB utilizes various sources of data, such as wearable technologies, social media, and other digital platforms, to gather information on customer behaviors and preferences. This data is then analyzed to derive meaningful insights that can drive decision-making and personalized experiences. The goal of IoB is to understand patterns, trends, and correlations in human behavior, which can be used in various domains such as marketing, healthcare, and public safety. By leveraging the power of data analytics and technology, IoB has the potential to enhance decision-making processes, improve customer experiences, and enable targeted interventions.

However, as with any technology that involves the collection and analysis of personal data, privacy, and ethical considerations are important. The responsible use of data and ensuring individuals' consent and privacy rights are crucial aspects of implementing IoB initiatives. The IoB is a concept that focuses on collecting and analyzing data to gain insights into human behaviors. It has the potential to revolutionize various industries by enabling personalized experiences and informed decision-making and these details are presented and discussed in fifth chapter of this book.

Cognitive computing is a field that involves the use of computerized models to simulate human thought processes in complex situations. It encompasses technologies influenced by cognitive science and artificial intelligence to mimic human thinking and decision-making. The goal of cognitive computing is to enable machines to understand, reason, learn, and interact with humans in a more natural and intelligent manner. It involves technologies such as machine learning, natural language processing, and data analytics to process and analyze large amounts of data and derive meaningful insights.

Cognitive computing systems [1, 2] are designed to handle ambiguous and unstructured information, making them well-suited for tasks that involve complex decision-making, problem-solving, and pattern recognition. They can assist in various domains, including healthcare, finance, customer service, and research.

Advancements in cognitive computing have led to the development of technologies like intelligent virtual assistants, behavioral analysis, biometrics, and automation. These technologies have the potential to enhance productivity, efficiency, and decision-making in various industries. However, cognitive computing also raises ethical considerations, such as privacy and bias, as these systems rely on large amounts of data and algorithms to make decisions. Responsible use and transparency are important aspects of implementing cognitive computing solutions and are discussed in the fifth chapter of this book.

The sixth chapter of this book deals with various business cases which includes Internet of Things, Healthcare, Musicology, Agriculture, and Interdisciplinary areas.

7.2 Current Trends and Challenges of Cognitive Computing for Machine Thinking

This section presents and discusses the current trends and challenges of cognitive computing for machine thinking.

a. **Current Trends**

As stated earlier, it is important to understand the recent trends in machine thinking [3]. Deep learning is a subset of machine learning that focuses on training artificial neural networks with multiple layers to recognize patterns and make predictions. It has been a major driver of recent advancements in AI, particularly in image and speech recognition, natural language processing, and autonomous systems. As AI systems become more complex and autonomous, there is a growing need for transparency and interpretability. Explainable AI aims to develop techniques and algorithms that can provide understandable explanations and reasoning behind AI decisions, making the decision-making process more transparent and accountable. Reinforcement learning is a type of machine learning that involves an agent learning to make decisions by interacting with an environment and receiving feedback in the form of rewards or penalties. It has gained significant attention in recent years, particularly in applications such as robotics, game playing, and autonomous vehicles.

Transfer learning involves training a machine learning model on one task and then applying the learned knowledge to a different but related task. It allows models to leverage pre-existing knowledge and speeds up the learning process, making it more efficient and effective in scenarios where large amounts of labeled data are not readily available. Generative models are AI models that can generate new data samples that resemble training data. This includes techniques like generative adversarial networks (GANs) and variational autoencoders (VAEs). Generative models have applications in image synthesis, text generation, and even creating deep fake content. Edge computing involves processing and analyzing data closer to the source or device, rather than relying on centralized cloud computing. This trend is driven by the need for real-time and low-latency processing, particularly in applications such as IoT devices, autonomous vehicles, and smart cities.

b. **Challenges**

Human thinking often deals with ambiguous and uncertain situations, where there may be multiple interpretations or incomplete information. Cognitive computing systems need to handle such complexity and develop mechanisms to reason and make decisions in such scenarios. Understanding natural language is a complex task. While computers are faster than humans at processing and calculating, they have yet to master tasks like comprehending the nuances and context of human language. Cognitive computing systems need to improve their natural language processing capabilities to effectively communicate and understand human language.

Human thinking involves acquiring, organizing, and applying knowledge. Cognitive computing systems need to develop effective methods for representing and learning from vast amounts of data and knowledge. This includes capturing and updating knowledge dynamically, handling evolving information, and adapting to new situations. Human thinking often relies on understanding the context in which information is presented. Cognitive computing systems need to develop contextual understanding capabilities to interpret and analyze information in a way that aligns with human thinking. This includes understanding the broader context, background knowledge, and situational factors that influence decision-making.

Cognitive computing systems need to address ethical considerations and potential biases. Ensuring fairness, transparency, and accountability in decision-making processes is crucial. Additionally, considering the social implications of machine thinking, such as job displacement and inequality, is important in developing responsible and inclusive cognitive computing systems. Cognitive computing heavily relies on data for learning and decision-making. Ensuring the quality, accuracy, and representativeness of the data is essential. Data biases and skewed datasets can lead to biased outcomes and reinforce existing inequalities. Cognitive computing systems need to address these challenges to provide fair and unbiased results. Simulating human thought processes requires significant computational power and resources. Cognitive computing systems need to handle the computational complexity of processing large amounts of data, running sophisticated algorithms, and performing real-time decision-making.

7.3 Ethical and Social Implications

This section presents and discusses ethical [4] and social implications [5] of cognitive computing for machine thinking.

a. **Ethical Implications**

With the increasing impact of AI on society, there is a growing focus on ensuring ethical practices in machine thinking. This includes considerations such as fairness, transparency, accountability, and bias mitigation in AI algorithms and decision-making processes. Machine learning algorithms can inadvertently perpetuate biases present in the data they are trained on, leading to discriminatory outcomes. This can result in unfair treatment in areas such as hiring, lending, and law enforcement. Addressing bias and ensuring fairness in AI systems is a critical ethical consideration. Machine thinking often relies on large amounts of data, raising concerns about the privacy and security of personal information. Ethical considerations involve ensuring proper consent, data anonymization, and secure handling of sensitive data to protect individuals' privacy rights.

AI systems can make decisions or recommendations that have significant impacts on individuals or society. Ensuring accountability and transparency in how these decisions are made is crucial. Understanding the inner workings of AI models, providing explanations for decisions, and having mechanisms for redress are important ethical considerations. As AI systems become more autonomous, questions arise about the level of human control and oversight. The ethical implications involve determining the appropriate balance between human decision-making and machine autonomy, particularly in critical domains like healthcare, transportation, and warfare.

Machine thinking and automation can lead to job displacement and economic inequality. Ethical considerations include addressing the impact on workers, ensuring retraining opportunities, and creating policies to mitigate socioeconomic disparities. The ability of AI systems to generate realistic content raises concerns about the manipulation of information and the creation of deep fake content. Ethical implications involve ensuring the responsible use of AI technologies to prevent misinformation, deception, and malicious activities.

The computational and energy requirements of machine thinking can have environmental consequences. Ethical considerations involve developing energy-efficient algorithms and promoting sustainable practices in AI research and development. Addressing these ethical implications requires a multidisciplinary approach involving technologists, policymakers, ethicists, and society at large. It involves developing ethical frameworks, regulations, and responsible practices to ensure that machine thinking aligns with human values, societal well-being, and the public interest. With the advancement of machine thinking, there are important ethical and social implications to consider. Researchers are examining the impact of machine thinking on privacy, bias, fairness, and accountability, and exploring ways to ensure responsible and ethical use of these technologies. Ethical decision-making by machine decides how machines can be programmed to make ethical decisions and navigate moral dilemmas. This research area explores the development of ethical frameworks and algorithms that guide machines in ethical reasoning. Bias and fairness in machine thinking examine the potential biases in machine thinking and the implications for fairness in decision-making. This research area focuses on identifying and mitigating biases in algorithms and ensuring fairness in outcomes.

Transparency and explainability refers to studying how to make machine thinking more transparent and understandable to humans. This research area explores techniques to provide explanations for machine decisions and increase transparency in the decision-making process. Accountability and responsibility help to investigate the allocation of accountability and responsibility in machine thinking. This research area examines legal and ethical frameworks for holding machines accountable for their actions and determining liability in case of errors or harm.

Privacy and data ethics examines the ethical implications of machine thinking in relation to privacy and data protection. This research area explores issues such as data collection, storage, and usage, as well as the potential for surveillance and intrusion into individuals' privacy. Social impact and human well-being analyze the broader social impact of machine thinking on individuals and society. This research area explores the effects of automation on employment, economic inequality, social interactions, and overall human well-being. Ethics in autonomous systems studies the ethical considerations in the deployment of autonomous systems powered by machine thinking, such as self-driving cars or autonomous drones. This research area addresses issues like safety, risk assessment, and the ethical implications of delegating decision-making to machines.

Human–machine collaboration investigates how machines and humans can work together in a way that respects ethical principles and promotes mutual understanding. This research area explores the dynamics of collaboration, shared decision-making, and the ethical implications of human–machine interactions.

b. **Social Implications**

Machine thinking and AI have the potential to automate various jobs and tasks, leading to concerns about job displacement and unemployment. This can have wide-ranging social and economic consequences, requiring the development of policies to address the impact on workers and promote job retraining. The adoption and deployment of machine thinking technologies may exacerbate existing social inequalities. Access to AI tools, resources, and benefits may be unevenly distributed, leading to a digital divide. Efforts should be made to ensure equitable access to AI technologies and bridge the gap between different communities.

Machine thinking can transform the delivery of social services and public sector operations. It has the potential to improve efficiency, effectiveness, and accessibility, but it also raises questions about privacy, data security, and the potential for algorithmic biases. Ethical considerations must be considered when implementing AI in public services. AI-powered systems can shape people's opinions, behaviors, and preferences. This raises concerns about the potential for manipulation and the impact on democratic processes, personal autonomy, and individual agency. Transparency, accountability, and safeguards are essential to mitigate these risks.

Machine thinking is reshaping the skills required in the workforce. It is important to prepare individuals with the necessary digital literacy, critical thinking, and problem-solving skills to navigate the AI-driven society. Educational systems need to adapt to ensure individuals are equipped to understand, interact with, and benefit from AI technologies. The rise of AI-powered systems, such as chatbots and virtual assistants, can impact human interactions and relationships. It is important to consider the potential consequences on social dynamics, interpersonal communication, and emotional well-being as AI becomes more prevalent in various aspects of daily life. Machine thinking raises complex ethical and legal questions that need to be addressed to protect societal values and individual rights. Issues such as privacy, bias, accountability, and fairness in decision-making should be carefully considered when developing AI systems.

7.4 Research Openings

There is ongoing debate about whether machines can truly think like humans, especially when it comes to consciousness and sentience. The definition of thinking in the context of machines is often tied to processing information, making decisions, and problem-solving. Machine thinking is a field which involves studying and developing machines that can mimic human thought processes and make decisions. Although the field of machine thinking is not yet explored completely, there are many areas of machine thinking which need extensive research [6]. Few of the research areas are listed below:

1. **Artificial Intelligence (AI) and Machine Learning**

AI and machine learning are key areas of research when it comes to machine thinking. Researchers are exploring various algorithms and techniques to enable machines to learn, reason, and make decisions in a way that resembles human thinking.

There are research agendas proposed for investigating computational thinking and the notional machine, particularly in the field of pre-service education and training for teachers. The future of machine intelligence and its ability to think like humans depends on advancements in understanding consciousness and developing more powerful and adaptive machines. Machine thinking is an area of interest and exploration in fields such as robotics, automation, and AI, with potential applications in various domains, including healthcare, finance, and scientific research. Machines can process information, make decisions, and perform repetitive tasks with automation technologies like machine learning and robotics. However, there are skills that machines currently struggle with, such as critical thinking, creativity, synthesizing, problem-solving, and innovating, which are often associated with human capabilities. There are various research openings in machine thinking, including exploring the boundaries of machine intelligence, understanding the limitations of machines, and investigating the potential applications of machine thinking in different fields. Researchers can delve into topics such as consciousness, decision-making algorithms, human–machine interaction, and the impact of machine thinking on society and the workforce.

2. **Cognitive Computing**

Cognitive computing is an interdisciplinary field that combines elements of AI, neuroscience, psychology, and linguistics to create systems that can mimic human cognitive processes. Researchers are investigating ways to develop machines that can understand and respond to natural language, recognize patterns, and solve complex problems. Natural language processing focuses on developing algorithms and models that enable machines to understand and generate human language. This includes tasks such as language translation, sentiment analysis, and chatbot development. Machine learning plays a crucial role in cognitive computing by enabling machines to learn from data and make predictions or decisions. Research in this area involves developing advanced machine learning algorithms and techniques to improve the accuracy and efficiency of cognitive systems.

Cognitive computing systems need to represent knowledge in a structured manner and use reasoning mechanisms to make intelligent decisions. Research in this area involves developing knowledge representation frameworks, reasoning algorithms, and knowledge-based systems. Computer vision is an important component of cognitive computing, enabling machines to understand and interpret visual information. Research in this area focuses on developing algorithms for object recognition, image classification, and scene understanding. Understanding human emotions and sentiments is crucial for cognitive systems to interact effectively with users. Research in this area involves developing models and techniques for emotion detection, sentiment analysis, and affective computing.

3. **Explainable AI**

Explainable AI (XAI) aims to develop machine learning models and algorithms that can provide transparent and interpretable explanations for their decisions and actions. This research area is important for building trust in AI systems and ensuring accountability. XAI is an emerging AI technique which involves AI outcomes with explanations and interpretability. Developing algorithms and techniques to understand and interpret the decisions made by AI systems is a key challenge. This involves designing models that provide insights into the reasoning process and factors that contribute to the decision-making. Investigating ways to improve the transparency of AI systems, enabling users to understand why specific decisions were made is another important challenge. This research area aims to build trust and accountability in AI by providing explanations that are understandable and meaningful to users. Examining the ethical implications of AI systems and ensuring that XAI methods adhere to ethical principles is extremely important. This research area focuses on addressing biases, fairness, and accountability in AI decision-making processes.

Enhancing the interaction between humans and AI systems by developing user-friendly interfaces that provide explanations for AI decisions is a need of hour. This research area aims to bridge the gap between AI algorithms and human understanding, making AI systems more accessible and usable. In addition to this, defining metrics and benchmarks for evaluating the explainability of AI systems is very crucial. This research area focuses on developing standardized evaluation methods to assess the quality and effectiveness of XAI techniques. Studying how humans perceive and understand explanations provided by AI systems is referred to as human perception of XAI. This research area investigates the cognitive processes involved in comprehending AI decisions and explores ways to improve the interpretability of XAI methods. Exploring techniques to make deep learning models more explainable is another emerging research area. This research area focuses on developing methods to extract meaningful explanations from complex deep learning architectures.

4. **Human–Machine Interaction**

Researchers are studying how humans and machines can effectively interact and collaborate in decision-making processes. This includes exploring ways to design user interfaces and systems that can understand and interpret human intentions and preferences, as well as techniques for humans to understand and trust the decisions made by machines.

Usability and user experience deals with studying how users interact with machines and designing interfaces that are intuitive, efficient, and enjoyable to use. This research area focuses on enhancing the overall user experience and optimizing the usability of machines. Natural language processing and dialogue systems develop techniques and algorithms to enable machines to understand and generate natural language, facilitating more effective communication between humans and machines. This research area includes areas such as speech recognition, natural language understanding, and dialogue management. Gesture and multimodal interaction investigate how humans can interact with machines using gestures, body movements, and other forms of non-verbal communication. This research area explores ways to design interfaces that can interpret and respond to these interactions effectively.

Exploring the use of AR and VR technologies to enhance human–machine interaction is another research area. This research area focuses on creating immersive and interactive experiences that blend the physical and virtual worlds. Human–robot interaction investigates how humans and robots can interact and collaborate in various domains. This research area explores ways to design robot behaviors and interfaces that enable effective communication and cooperation between humans and robots. Accessible and inclusive design deals with how to design machines and interfaces that are accessible to a diverse range of users, including those with disabilities or special needs. This research area focuses on ensuring equal access and usability for all individuals.

Cognitive modeling and human factors investigate human cognitive processes and factors that impact human–machine interaction. This research area aims to understand how humans perceive, process, and make decisions in interaction with machines, and how to design interfaces that align with human cognitive capabilities.

7.5 Future Outlook

The outlook of machine thinking is promising and has significant implications for various industries and society. The potential areas are listed below:

1. **Advancements in AI**: IBM has been at the forefront of advancing AI-driven technologies, which indicates a continued focus on developing machine thinking capabilities.
2. **Future of Jobs**: The World Economic Forum's Future of Jobs report highlights the shifting human–machine frontier and the evolving nature of work due to the integration of machines and AI technologies.

3. **Career Opportunities**: Machine learning concepts like computer vision are opening exciting career opportunities for forward-thinking individuals, according to a report on AI and machine learning job trends.
4. **Impact on Jobs**: There are discussions about the impact of machines on human jobs, with concerns about job displacement and the emergence of new educational and training needs.
5. **Capabilities versus Human Skills**: While machines and algorithms have advanced capabilities, there are still areas where humans excel, such as services that require complex thinking and human interaction.
6. **Automation and Workforce**: As machines increasingly complement human labor, there is a need to address challenges and solve issues related to automation, beyond-human capabilities, and computer vision.
7. **Positive Job Outlook**: The job outlook for computer and information research scientists, who work on machine thinking technologies, is projected to grow significantly in the coming years.

This indicates that machine thinking will continue to advance and have a substantial impact on various aspects of society, including the job market, career opportunities, and the nature of work. However, it is also important to consider the ethical and social implications associated with the integration of machines and AI technologies to ensure responsible and beneficial adoption.

Machine thinking application development is complex and computationally expensive. Technology requirements of machine thinking includes:

- **Computing Power**: Machine thinking often requires significant computing power to process large datasets and run complex algorithms. High-performance hardware, such as powerful CPUs or GPUs, may be necessary to handle computational demands.
- **Data Storage and Management**: Machine thinking relies on vast amounts of data. Adequate storage infrastructure is crucial to store and manage the data effectively. This may involve scalable storage solutions like cloud storage or distributed file systems.
- **Networking and Connectivity**: Machine thinking applications often require fast and reliable network connectivity to access data sources, communicate with other systems, or leverage cloud-based services. Robust networking infrastructure is essential for seamless data transfer and real-time communication.
- **Software and Development Tools**: Machine thinking typically involves using specialized software frameworks and development tools. Popular machine learning frameworks like TensorFlow or PyTorch are often used to build and train models. Additionally, programming languages like Python and R are commonly employed in machine thinking projects.
- **Data Processing and Analytics**: Machine thinking often involves preprocessing and analyzing data. This may require software tools and libraries for data cleaning, feature extraction, and exploratory data analysis. Big data processing frameworks like Apache Spark can be utilized for handling large-scale data processing tasks.

- **Model Deployment and Integration**: Once machine thinking models are developed, they need to be deployed and integrated into existing systems or applications. This may involve creating APIs or services to expose the models for real-time predictions or integrating them into larger software architectures.
- **Security and Privacy**: Machine thinking systems often deal with sensitive data, so ensuring security and privacy is crucial. Robust security measures, such as encryption, access controls, and data anonymization techniques, should be implemented to protect the data and models.
- **Monitoring and Maintenance**: Machine thinking systems require ongoing monitoring and maintenance to ensure optimal performance. This may involve setting up monitoring tools, tracking model performance, and addressing issues like model drift or concept drift over time.

It's important to note that the specific technology requirements for machine thinking can vary widely depending on the application and use case. Organizations and practitioners should carefully consider their unique needs and consult domain experts to determine the most suitable technology stack for their machine thinking projects.

References

1. Taylor JG (2009) Cognitive computation. Cogn Comput 1:4–16
2. Chen M, Herrera F, Hwang K (2018) Cognitive computing: architecture, technologies, and intelligent applications. IEEE Access 6:19774–19783
3. Qiu J, Wu Q, Ding G, Xu Y, Feng S (2016) A survey of machine learning for big data processing. EURASIP J Adv Signal Process. 2016(1):1–16
4. Char DS, Abràmoff MD, Feudtner C (2020) Identifying ethical considerations for machine learning healthcare applications. Am J Bioeth 20(11):7–17
5. Borrellas P, Unceta I (2021) The challenges of machine learning and their economic implications. Entropy 23:275
6. Zhou ZH (2016) Learnware: on the future of machine learning. Front Comput Sci 10:589–590

The manufacturer's authorised representative in the EU is Springer Nature Customer Service Centre GmbH, Europaplatz 3, 69115 Heidelberg, Germany. If you have any concerns regarding our products, please contact ProductSafety@springernature.com

Printed and bound by CPI Group (UK) Ltd, Croydon, CR0 4YY
01/10/2026
02225947-0001